I0438455

TEN
MINUTE
ECOLOGIST

Twenty Answered Questions for Busy
People Facing Environmental Issues

John Janovy, Jr.   2nd Edition

Ten Minute Ecologist, Second Edition. Copyright © 2011 by John Janovy, Jr. All rights reserved. Printed in the United States of America. No part of this book may be used or reproduced in any manner whatsoever without written permission except in the case of brief quotations (600 words or fewer) in scholarly publications, articles, or reviews. For additional information contact the author at jjparasite@hotmail.com.

Designed by John Janovy, Jr.

ISBN-10: 1466393998
ISBN-13: 978-1466393998

Books by John Janovy, Jr.

Keith County Journal

Yellowlegs

Back in Keith County

On Becoming a Biologist

Fields of Friendly Strife

Vermilion Sea

Dunwoody Pond

Biodiversity: A Primer (with Amanda Snyder)

Comes the Millennium (as Jack Blake)

Foundations of Parasitology (with Larry Roberts)

Ten Minute Ecologist

Teaching in Eden

Outwitting College Professors

The Ginkgo

Conversations between God and Satan

Pieces of the Plains

Tuskers

Intelligent Designer: Evolution for Politicians

TEN MINUTE ECOLOGIST

Contents:

Foreword — 7

1. How do we humans view the world? — 9
2. What is a species? — 15
3. What is biodiversity? — 21
4. What is dirt? — 29
5. What is water? — 35
6. What is air? — 41
7. Who eats whom? — 49
8. Who beats whom? — 57
9. What is an ecosystem? — 63
10. Why are the tropics so complicated? — 69
11. Why is the Arctic so fragile? — 77
12. Why study islands? — 83
13. How is real estate really divided up? — 89
14. How many is too many? — 95
15. How long is short? (or, How short is long?) — 101
16. What good is a swamp? — 107
17. Why are ecologists such nerds? — 115
18. Why do scientists argue? — 121
19. Do we humans live by the same rules as beetles? — 127
20. Did God make the Earth in seven days? — 133

Glossary — 139

Acknowledgments — 151

Suggested readings — 153

About the Author — 161

Foreword

I was sitting in a meeting one day listening to one of the world's most distinguished scientists talk about biodiversity. His audience was made up mainly of business executives and attorneys who, because of factors such as government regulations or marketplace events, suddenly found themselves dealing with environmental issues. As I looked around the room, I could see the audience paying close attention to the speaker. But afterwards someone said to me: I loved that speech but I still don't know what "biodiversity" really means or why it's so important. At that point I decided all these businessmen needed help, but they didn't have the time to go back to college and major in biology. That's when I decided to write this book.

Ten Minute Ecologist was originally intended for business executives suddenly in need of education on environmental matters, whether because of regulations, public protests, or even a feeling that maybe they should be doing "something for the environment." It also is an excellent book for anyone newly involved in ecological issues, regardless of the nature of that involvement. My goal is to avoid polemics and to provide plain, easily understood, non-threatening, information. I'm not taking a side in any debate, unless it's the evolution-creation one, which is not really a debate so much as a political phenomenon stemming from willful ignorance. Nor do I believe this book is all a person needs in order to interpret the positions in a controversy.

Ten Minute Ecologist will do wonders for anyone who is now a CEO on his way to a public hearing on an environmental impact statement but can't remember anything from 10th grade biology. It also will do wonders for anyone else who does not consider himself or herself a trained ecologist, but who reads about environmental issues in the daily newspapers and asks "should I be worried about this?" My early manuscript readers were public school teachers; their favorable response to the "work in progress" suggests that *Ten Minute Ecologist* may also find a useful place on teachers' bookshelves. Regardless of who ends up reading it, I hope that *TME* actually serves as an introduction to the appendix—a list of suggested books that build on some of the themes presented here.

I wrote *Ten Minute Ecologist* as a series of answers to twenty questions. Each answer should take about ten minutes to read, thus the title. Nobody's going to become an ecologist on ten minutes a day. But if we don't all make an effort to become more scientifically literate, then environmental issues will continue to be resolved in an atmosphere of public ignorance. There are not many secrets or big discoveries revealed in the following pages. Virtually all of the information in this book is fairly common knowledge to professional ecologists and many of the examples I've used have been written about time after time and cited in all leading texts.

But unlike scholarly textbooks, this one contains answered questions, questions that I've heard asked again and again by responsible, concerned, well-educated and successful people who are not in a position to go back to college and major in biology. I've often felt that if paperback fiction, movies, and television, all dealt with substantive ecological issues at the same level they deal with issues of politics, law, race, and economics, then maybe as a nation we'd be better educated on the way natural systems operate. As a result of this feeling, I've tried very hard to make reasonably complex ideas accessible to the same audience that reads paperbacks and watches television.

John Janovy, Jr.
October, 2011

1. How do we humans view the world?

Vision is the art of seeing things in-visible.

—Jonathan Swift (from
*Thoughts on Various
Subjects; from Misce-
llanies*, 1711)

Humans are large, short sighted, very intelligent, an-imals. This set of traits dictates how we view the world. Our world view, in turn, typically directs our actions. Actions, whether by humans or insects, always have an effect on an organism's immediate environment. As a minimum, we all consume food and produce wastes; many species, including our own, use natural materials to build "houses." We re-produce and increase in numbers (actions), and we alter our environments in order to accomplish this reproduction (effects). Other species influence our lives (disease causing organisms, food and fiber producing organisms); an obser-vant trip to the grocery store will reveal the extent to which this claim is true.

Non-living elements of our environment—e.g., tem-perature, water, minerals—also influence our actions. Organ-isms and their products are called biotic factors of the envir-onment; non-living elements are abiotic factors. Ecology is the study of the interactions between organisms, for example humans, and both the abiotic and biotic factors that influence their lives. Ecological interactions are dictated to a great ex-tent by a species' inherited traits, and for humans, "inherited traits" means, as a minimum, large size, short-sightedness, and intelligence.

Our large size is one of the most important of our traits, important in the sense that it strongly influences our interactions with other species. We see other large animals, and large plants, before we notice smaller ones, and typically we need to learn how to look for small organisms before we notice them at all. The smaller the plants and animals, the worse the problem of non-recognition. But the vast majority of living things are tiny. The average size of animals is about half an inch long. Thus we are predisposed, by virtue of having been born human, to ignore most of the living creatures whose lives also depend on the only planet in the universe known to support life.

This problem is the major one most people must solve when trying to understand the natural world. Not only a ten minute ecologist, but a ten year ecologist as well, must accept the fact that our view is distorted by our size then learn to "see" the unseeable. Tiny worms occur by the millions in many if not most lakes and many streams; bacteria cover the surface of every grain of sand; microscopic plants fill the oceans; fungi "eat" the beautiful fall leaves that finally end up on the forest floor.

Is all this little life important? Yes, it is very important. The richness of the soil, the ability of plants to trap nitrogen and make protein, and a host of other fundamental life processes depend on microscopic organisms. Most important of these processes is movement of nitrogen through the environment. Without microscopic organisms, especially bacteria associated with plant roots, we would not be able to utilize the nitrogen in the air. Without nitrogen, we cannot build the proteins that are an essential part of our bodies. Without nitrogen, we cannot build the genetic information we pass on to our offspring. These statements are true for snakes and fish as well as for humans. In later chapters the details of nutrient cycles and energy flow are discussed. But for now, we need only to remember that the flow of nutrients into our food, thence into our bodies, is ultimately dependent on microscopic life in the soil, water, and air.

Our short sightedness is sometimes considered a trait that we inherited from our non-human ancestors. Indeed, short sightedness is far and away the most common shared trait among animals. That is, virtually all species respond to immediate environmental changes (shifting light patterns, noises, passing members of the opposite sex, etc.) and tend to ignore the more "distant" factors such as next year's price of corn, predicted oil shortages, and the like. We are not much different from the grasshoppers in this regard. Although certain individuals, and the insurance industry, try to plan far ahead, most of us spend most of our time worrying about and reacting to events that occur over an hour, a day or a week.

Language and culture have given humans the power to act over great distances and relatively long periods of time, however. Furthermore, culture provides us with the power to act in large groups—e.g., nations and industries—to accomplish large tasks we could not do either as individuals or as families. Language and culture are as much a part of our basic biology as our short-sightedness. Thus we've inherited two opposing traits, one from our non-human ancestors (short-sightedness), and the other from our human ancestors (language and culture). One of these traits gives us power to act over long distances and times; the other trait blinds us to the results of such acts.

"Environmental problems" are mainly human problems produced by the interaction between these two opposing traits. For example, because of our short-sightedness, we consume resources without worrying much about the long term supply of them (water, fossil fuel). But, because of our language and culture, i.e. our humanness, we are disturbed by the thought of destroying the natural world, which many of us view as God's creation.

Plants and animals, including entire species that may be displaced by our actions, don't worry about anything; they just live or die. Aside from possibly apes and whales, no other species have language and culture in the human sense. So "environmental problems" have a very real philosophical component, namely the question of how humans should interact with their environments, with the other species that occupy Earth, and with each other as individuals. But "environmental problems" also have a practical component because Earth's ability to support human life depends to a certain extent on the planet's supply of non-human organisms, and on Earth's ability to sustain that supply.

Language and culture are products of our intelligence, and in turn, actions made possible by our intelligence are modified by language and culture. Thus we also have a positive feed-back system. Computers are a superb example of this feedback system at work. Mainly during the 20th Century, and primarily during the last half of that century, our intelligence gave us the power and inclination to think about ways to carry out certain calculations, and mathematical procedures, faster and more accurately than could be done by people with pencils. One remarkable result of these language and culture activities (thinking and building) was the computer. The computer has, in turn, significantly modified our language and culture, providing new metaphors, new kinds of interactions between individuals and nations, new ways of waging war, and new sources of information.

The computer is only one of a great many excellent examples of human products that alter our vision, and which are then themselves altered by that vision. Indeed, many, and possibly most, humans tend to view the world in terms of human products and through these products' vision-altering effects. Human history is short, in the natural sense; a few thousand years is not a very long time. But during that few thousand years, we've been building things, modifying our language and culture, and consequently changing our view of the world because of what we've built. Some people claim that because of these feedback cycles humans have lost touch with the planet that supports them. Other people claim that human activities are perfectly natural because we are, after all, a species of living organism that evolved on Earth. When these two groups of people get together, the result is often conflict over an environmental issue.

In the not too distant past, human products were primarily art, music, literature, machines and buildings, i.e., relatively long lived tangible items. Recently, however, human products have come to include electronic images and sounds, which bear only superficial resemblance to anything alive. These kinds of images and sounds can be manipulated in uncountable ways, and the results are often quite seductive for humans which are, by definition, very intelligent animals that use their imaginations and often seek mental stimulation. We have thus generated a philosophical problem, namely the question of whether our electronic images are "real." The answer is relatively simple: yes. Images are real. But (and this is a fairly large "but") these very real images, real in the philosophical sense, may have absolutely no connection with any of the natural processes that support life on Earth, and indeed may convince us that we've successfully separated ourselves from these processes.

Thus we've arrived at perhaps the most central issue of humanity's modern world view, namely the question of whether our perceptions, upon which we so often act, match the reality of having been born large, short-sighted, and ultimately dependent on processes carried out by bacteria. The answer to this question varies according to individuals, societies, and periods of history. The deeply religious among us may claim that the answer does not matter, for our lives are governed by forces over which we have no control, committed to ends we cannot fathom. The more cynical among us may claim that in an electronic age we lack the capacity for joining our perception with the more Earthly kinds of reality.

Between these two extremes lies a vast territory of human thought, action, and rationale. Somewhere in that territory stands the ecologist who believes that humans must act wisely in their relationships with Earth. And the more tightly human perception is governed by knowledge of the planet, the ecologist claims, the better the future looks for all of us.

2. What is a species?

*. . . and whatever the man called a living
creature, that was its name.*
—Genesis 2:19

A species is a kind of plant, animal, fungus, or microbe. The word "species" is both singular and plural (a species, many species). Most people can distinguish a few common kinds (species)—dog, cat, rose. Some people, particularly hunters and fishermen, can distinguish a few more species—quail, pheasant, mallard, pintail, walleye, bluegill, channel catfish. People who work in greenhouses and nurseries usually learn to distinguish several dozen species. Habitual zoo goers learn some exotic ones—gorilla, hyena, zebra, crowned crane. In the overwhelming majority of cases, the species recognized by an average citizen are, like the citizen, large ones. Most people are quite surprised to learn that what they think of as one species, may actually be several.

"What is a species?" is one of the most persistent questions in biology. People have been asking it for a very long time, and arguing about their answers for an equally long time. The question is persistent, and the answers open to discussion, because the criteria for distinguishing kinds are not always obvious. It's not too difficult to tell a zebra from a quail. But it's a little more of a challenge to learn to distinguish the three different species of zebras from one another, or the six or seven species of quail that live in the United States. In addition, remember, the vast majority of species are small, and belong to groups of plants and animals that we think of as exotic. In addition, untold numbers of microbial species remain to be discovered. Small size and lack of information also contribute to our uncertainty about what constitutes a species among these relatively unknown organisms.

Most biologists use structural characteristics to describe species of organisms we call plants, animals, and fungi. Microbiologists may use some functional characteristics to distinguish between bacteria, e.g. ability to grow in a certain kind of broth. Modern science has produced numerous biochemical methods for distinguishing between kinds, e.g., by use of DNA (raw genetic information), but it's still not always clear how best to use such methods.

In current discussions of ecology, especially if politics are involved, one often hears about "known species," "unknown species," and "endangered species." I'll deal with these subjects in a moment, but before I do, there are two additional things one needs to know about species. First, they are generally understood (assumed, considered) to be reproductively isolated from one another. That is, in most cases, two different species either cannot or do not interbreed and produce fertile offspring. Second, the species is considered to be an evolving unit. That is, species (reproductively isolated kinds) evolve from other species and eventually evolve into still other species. Species also disappear from the face of Earth, a process known as extinction. A species that dies out, down to the very last individual, has become extinct. So has a species that's evolved into another one.

"Known" species are better called "described" species. By "described," we mean that someone has taken the trouble to study the organism very carefully, decided which structures are best used to distinguish it from others, written a paper containing the results of such study, and published this paper in a scholarly publication so that other scientists have access to the information. Such publication means that the world now knows about the existence of this species, thus it is a "known" one, as opposed to a yet-to-be-discovered one.

In many cases, and ideally, specimens of the new species are deposited in a museum. Thus if a scientist collects some organisms, say ants, and wants to know what kind they are, then he or she consults the published scholarly works on ants in order to learn how to identify the specimens, and may actually borrow museum specimens to use in accomplishing the task. For this reason, museum research collections are exceedingly valuable, and the expertise needed to maintain such collections helps dispel our ignorance about the natural world.

Attempts to identify specimens sometimes end in failure. That is, you simply cannot determine, from the scientific literature, what kind of organism you have in hand. At this point, you become suspicious that you're looking at a new species that must be "described" as above. You then do hundreds of measurements, make dozens of drawings, compare structures with those of known species, sometimes do growth studies or DNA analysis, write a paper summarizing the results of all this work, then submit this paper to a learned journal. Experts review the research, and if they find it acceptable, the paper gets published. You are then known in the discipline as the describer of a new species.

Described species must have scientific names assigned to them. Scientific names are much more than nerdy arcane professorly jargon; they are, instead, guides to the literature. Through use of a scientific name, a scientist can discover a massive amount of information about a species because that same name is used in all scientific literature about the species. Sometimes scientific names are changed because people do additional research and discover that they need to be changed for a variety of reasons too involved to be included in this book. However, someone skilled in the use of scientific names can trace these kinds of decisions backward in history to the original published description.

Scientific names of known plants, animals, fungi, and bacteria consist of two words: a genus name and a specific name (the latter is technically known as a "specific epithet"). For example, the familiar domestic dog is *Canis familiaris*. *Canis* is the generic name; *familiaris* is the specific name (= specific epithet). This is the name given to the dog by Carl von Linné (Linnaeus), the man who invented this system of nomenclature in the 1700s. In print, scientific names are italicized.

Sometimes species are named in honor of people, e.g. *Salsuginus thalkeni*, a worm named after a landowner (Mr. Thalken) who gave scientists permission to use his property, or *Actinocephalus carrilynnae*, a species named after the describer's sister Carri Lynn. Species may also be named from some notable character they have, the place they were found, or other reasons. Usually the newly discovered species belongs to a known genus, which fact dictates the generic name. If it does not, then the scientist must describe a new genus, too, to go along with the new species, and make the case convincingly to anonymous editorial reviewers.

Unknown species are those living in nature, unstudied, in many cases undiscovered, and certainly undescribed, formally, by scientists. Most people are very surprised to learn how common unknown species are. They are everywhere. We have a long way to go before we discover all the species that occupy this planet. The fact that most organisms are small means that a great many evade discovery. The fact that most organisms live in the tropics, where field work is expensive and difficult, further contributes to their obscurity. In addition, human scientists are large animals that generally prefer to study large organisms, so tend to ignore the small ones unless these scientists are, for some unexplained reason, captivated by the microscopic world. Unknown species are so common that almost any serious undergraduate biology student can discover and describe one, even in the United States and Europe, sometimes in a local meadow, especially if that student is willing to learn how to use a microscope well.

One of the central issues in the current environmental debate concerns "biodiversity," a term that refers to numbers of species. The main reason biodiversity is an issue is this ease of finding unknown (= new, in the scientific sense) species. Thus scientists are able, by virtue of their knowledge about the number of unknown species they discover with a given amount of effort, to predict how many are yet to be discovered. From this kind of scientific activity, we know that our current inventory of known species, even though it numbers well over a million, is far from complete, and in fact may be but a small fraction of the kinds of organisms that actually occupy Earth with us. Scientists get uncomfortable in the presence of ignorance, especially when they can see decisions being made without enough knowledge to understand the consequences. This is the reason "biodiversity" is an environmental hot button.

"Endangered species" are those whose populations are so low they may not be able to survive. Every species needs to be present on Earth in certain numbers to ensure survival. That is, the individuals must be able to find mates and produce offspring, or else the species will become extinct. A species needs a certain kind of habitat—e.g. prairie, marsh, mountain top, desert—to live in. If the habitat disappears, then the species does too, and if enough of the habitat disappears, then the species becomes extinct. Species' populations also need to be high enough to maintain genetic diversity through interbreeding.

Although different species do not interbreed (in theory and usually in practice), individuals within a species (e.g., your friends) vary genetically. When genetic variation is low, especially in small populations, then the species becomes susceptible to the effects of inbreeding, mainly an increased frequency of genetic defects and increased susceptibility to disease. For this reason major zoos screen their

19

specimens genetically and try to exchange animals for breeding. As an aside, there are very many species, typically microorganisms, that do not reproduce sexually, at least all the time. In these cases, reproductive isolation cannot be the philosophical basis for species distinction. Instead, such organisms exist on Earth as populations of clones. But we still often measure them, draw pictures of them, analyze their DNA, and describe them as species.

Although the current system we have for naming organisms is based primarily on the 10th edition, in 1758, of Linneaus' *Systema Naturae*, modern molecular methods have produced some interesting modifications of this system. Knowledge of DNA structure sometimes alters our perception of who's related to whom, and thus which species truly belong together in a group. Taxonomy is the science of classification; systematics is the science of classification as it is applied to questions of evolutionary relationships. Like all sciences, taxonomy and systematics change with new technology, including that developed by the molecular biologists. Thus recent studies of relationships almost always involve analysis of genetic information, i.e., DNA.

Evolution is the central unifying theme of biology, and knowledge about the species present on Earth is central to our study and application of this theme. Much of what humans do with their biological resources, including plants with medicinal properties, species domesticated for agricultural uses, and species harvested from natural populations, depends on an accurate inventory of life on Earth. This inventory includes knowledge of the kinds of organisms present, the places these organisms live, and the sizes of the populations supported by their habitats, and it is very far from being complete. There are many scientists who believe there are so many species on Earth that no matter how hard we try, with the available talent and interest, we will never complete the inventory. Never.

3. What is biodiversity?

> *. . . approximately 835 genera and 20,000 to 25,000 species. . . a most bewildering diversity in the tropics . . .*
>
> *. . . with 650-660 genera, and about 10,000 species . . . cosmopolitan in distribution*
>
> —Dirk Walters and David Keil, referring to orchids and grasses, respectively (*Vascular Plant Taxoxomy*, 4th Ed., 1977)

The word "biodiversity" is a conjugation between the words "biological" and "diversity." "Biological" is an adjective that refers to life or life processes; "diversity" is a noun that means the state being diverse, or the condition of having different kinds of items. Because biological kinds are species, "biodiversity" therefore refers primarily to numbers of species. When a distinguished scientist uses the term biodiversity, he or she is often talking about the tropics, where most of the world's species live.

Humans are destroying these tropical forests at the rate of 50-100 acres a minute and have been for years. Because of what we know about the numbers of species in the tropics, this destruction means that we're also destroying most of the world's genetic information (= life on Earth) at the same rate we're destroying the forests. For this reason, biodiversity is an ecological issue among those who care about the world's supply of living organisms as well as among those who care nothing at all about the tropics, biological diversity, living organisms other than themselves or what's for dinner. The latter group just does not realize that such destruction of life on Earth will, ultimately, have unpleasant consequences for us all.

There are five general principles that ten minute ecologists need to know about biodiversity. These principles are:

(1) Most of the world's species live in the tropics;

(2) Prairies, wetlands, deserts, rocky coasts, deciduous forests, and other identifiable habitats, all have their characteristic fauna (animals) and flora (plants), as well as (insofar as we know) characteristic fungi and soil microbes. Biodiversity for a Nebraska prairie does not mean the same thing as biodiversity for a rocky coastline in Mexico.

(3) Most species live in and on other plants and animals; i.e. they are "symbionts" ("sym" = together; "bio" = life), a category that includes parasites. Every animal species that has been studied carefully has been shown to have one or more species of parasites; thus most of the world's animals are parasitic, and parasitism is therefore the most common way of life on Earth.

(4) Complex habitats support more species than do structurally simple habitats (I've devoted two chapters to this principle, chapter 10, "Why are the tropics so complicated?" and chapter 12, "Why study islands?"). "Complex" in this case refers mainly to the numbers of plant species; habitats with high numbers are complex. Complex places are usually varied in their physical attributes, too, for example an island with mountains, valleys, waterfalls, floodplains, and lakes. Such physical complexity almost always supports a relatively large variety of plants.

(5) All of our present agricultural plants and animals were once wild, and biologists feel that the tropical forests in particular are filled with species whose economic potential has never been discovered or developed.

In order to talk meaningfully about biodiversity, a person also has to understand something of the taxonomic (classification) system used to classify plants and animals. This system is hierarchical; that is, it is built of increasingly inclusive groups, which are called taxa (pl.; singular = taxon).

A genus (taxon), for example *Canis* (that of the familiar dog, Chapter 2), may contain several species. *Canis familiaris* is the dog, but *C. lupus* is the wolf and *C. latrans* is the coyote. So a genus includes at least one, and sometimes many, species, depending on the genus.

Genera (plural of genus) are grouped into more inclusive taxa, namely families, each containing at least one genus. Red foxes are similar to dogs, but not similar enough to be included in the same genus; the red fox (*Vulpes fulva*) is, however, in the same family as the dog (Canidae). Orders are still more inclusive, containing one or more families. Classes contain one or more orders. Phyla (pl.) are very inclusive taxa, and all of the members of a phylum (singular) have a shared basic body plan. Thus our familiar dog belongs to the species *Canis familiaris*, the genus *Canis* (with wolves and coyotes), the family Canidae (with foxes and African wild dogs), the order Carnivora (with bears, weasels, and skunks), the class Mammalia (with whales and humans), and the phylum Chordata (with fishes and frogs). Notice that in print, only the genus and specific names are italicized.

The more inclusive taxa (groups)—phyla, classes, and orders—are often called "higher taxa." The less inclusive taxa—genera, species—are called "lower taxa." Habitats can be relatively diverse because they contain many higher taxa (orders and classes), or because they contain many lower taxa (genera, species), or [usually] both. A detailed description of the many known taxa and their geographic distribution is well beyond the scope of this book. In fact, that inventory is so voluminous that it now fills the world's libraries. But, as I said in chapter 2, the inventory of biological diversity still is far from complete.

When an ecologist uses the term "diverse," he or she usually means "relatively diverse," although the term "rich" is also used commonly. The opposite of diverse (rich) is impoverished (poor). Some habitats are naturally impoverished (e.g. small islands); while others are naturally rich (e.g. large islands with tropical forests). In theory, it is easier to preserve the diversity of an area by preserving the habitat than to focus on the welfare of one or a few endangered species. In practice, however, preservation of habitat does not always ensure preservation of all the species that live in that habitat unless it is of a certain size. Thus a small marsh in the middle of a vast agricultural landscape does not necessarily support the same biodiversity as a large wetland in the middle of a vast expanse of ungrazed grasslands. This principle is explored in more depth in that chapter entitled "Why study islands?"

Ecologists have a number of formal ways to express biodiversity. These formal ways use mathematical equations that generate "diversity indices." In general, diversity indices (sing. = diversity index) take into account not only the number of species in a habitat, but also the proportionate distribution of those species. A group of interacting species sharing a habitat is called a community. Diversity of a community depends on two properties: the number of species present, and the evenness in number of individuals per species. A community in which all the species are about equally represented, numerically, is more diverse than one in which they are not. For example, if one community has three species—A, B, and C—with 100, 120, and 132 individuals respectively, then that community is more diverse than one with 175 As, 22 Bs, and 3 Cs. That is, when the math is finished, the diversity index of the first community will be higher than that of the second because the numbers of the species are closer to being equal.

But communities with relatively equal numbers of individuals in each species, e.g. as in the first example above, are not always the natural condition. Indeed, it is very unusual for numbers of individuals to be relatively equal. So diversity of natural systems (prairies, deserts, rocky seashores) is limited in part by natural processes that dictate the numbers and distributions of organisms. Ecologists have been very interested in these processes for a long time and have devoted much energy to discovering and describing them.

One of the ecologists' more interesting findings is that in most cases, some disturbance of a habitat is necessary before that habitat will support the maximum number of species that could live there. For example, unless prairies burn periodically, their biodiversity will diminish with time. The death of large trees (disturbance) helps maintain biodiversity of the biota (plants, animals, fungi, microbes). Why is some disturbance necessary to maintain biodiversity? The answer is: because some species are evolutionarily adapted to disturbed areas.

A second, and perhaps more general reason why some disturbance is often necessary to maintain a high number of species is that disturbance produces what ecologists call heterogeneity. "Hetero-" means different. A heterogeneous environment is one in which there are a lot of different places to live. The more heterogeneous the environment, the more places to live, and the more species one finds in that environment. This same principle applies to cities, too.

In addition to their distinctive structural features (see chapter 2), each species also has a fairly distinctive set of habitat (ecological) requirements. This set of requirements is called an ecological niche, and each species is said to occupy its unique ecological niche. Thus heterogeneous environments, with many places to live, also have many occupied niches, and consequently many species. Ecologists nowadays do not always agree on the best way to define, describe, interpret, and measure a species' ecological niche, and I've simplified this niche discussion somewhat, but it's not too far off base.

Scientists who are active in the conservation movement often deplore the "loss of biodiversity" due to human activities. These scientists are referring mostly to the clearing of natural habitats, e.g. prairies, wetlands, and forests, and conversion of these habitats into agricultural ones. A native prairie has many species of plants and animals; the cornfield made from native prairie has only one species (the farmer hopes), namely *Zea mays* (corn). Routinely the cornfield has other species, however, such as insects and weeds, which cannot be killed individually in high enough numbers to prevent them from harming the corn crop, so are killed collectively with pesticides.

In other cases, for example in the tropics, forests are burned, crops grown for a few years on poor soils, then cattle run on the land. The results of these tropical activities are different from the cleared prairie in kind but not in principle. Conversion of "natural" areas to agriculture results in homogeneity (sameness) where there was once heterogeneity (difference), thus loss of biodiversity. However, as long as there are increasingly large numbers of humans on Earth, natural habitats will be cleared for agriculture. Thus the same scientists who deplore the "loss of biodiversity" typically also deplore, not surprisingly, the exponential growth of human populations.

There are both philosophical and practical reasons for worrying about loss of global biodiversity. The philosophical reasons are explored in the last chapter ("Did God make the Earth in seven days?"); the practical reasons relate to the undiscovered medicinal properties and agricultural potential in the hundreds of thousands of organisms scientists have yet to study, and lessons to be learned about sustainable agriculture from research on natural habitats. But compared to some other attention-grabbing, short-term events such as car bombings in the Middle East, volcanoes, piracy off the Somalian coast, and terrorist attacks, loss of global diversity seems to be a rather mild, unimportant, tree-hugger, liberal sort of concern. Nothing could be further from the truth; loss of genetic information that spells "Life on Earth" is of staggering importance to the human species regardless of fact that it's happening over a lifetime instead of over an afternoon.

4. What is dirt?

This painted child of dirt, that stinks and stings;

—Alexander Pope (*An Essay on Man*, 1733-1734)

Dirt, or more properly soil, is an exceedingly complex mixture of organic (living, or derived from living) and inorganic (non-living) materials. Furthermore, the nature of this mixture varies enormously, sometimes even over short distances. Structure and organic content of dirt determine, in part, the makeup of the plant community that grows in it naturally. The other major factors in determining the makeup of the plant community are temperature and moisture, both of which influence dirt, too, but often indirectly. The natural plant community in an area reveals the agricultural potential of that area.

The combined effects of temperature and moisture, working over the long term to influence plant communities, are discussed in detail in chapter 13, entitled "How is real estate really divided up?" Although we're talking about dirt in these few pages, the makeup of dirt depends on a number of complex inter-actions between other living and non-living forces also discussed in later chapters. Because these inter-actions are often so complex, ecology—ten minute or otherwise—is not easily divided into chapters. But dirt's pretty basic to an understanding of ecology.

The elementary property of dirt is its "particle size distribution." Viewed under a microscope, dirt is seen to be made up of separate particles. These particles are what's left of mountains that have worn down from eons of being exposed to rain, freezing and thawing, blowing wind, running water, Earthquakes, volcanoes, rocks rubbing against other rocks, getting split by plant roots, and getting dissolved in acid. The particles eventually become dirt-sized and are transported, sometimes for great distances, mainly by wind and water.

If these particles are relatively large, we call the dirt "gravel." If particles are very small, and all about the same size, we call the dirt "clay." We have a lot of names for dirt with particle size in between gravel and clay. Some dirt is sand, some dirt is loam, some is sandy loam, and some is loess. The particles are not always, nor necessarily, all the same size, and certainly not the same shape. The "particle size distribution" is a listing of the various sizes of particles, and their relative proportions, in a sample of dirt. Many of dirt's other properties, e.g. its ability to hold water, and the kinds of organisms that live in this water, are related to its particle size distribution.

The second important property of dirt is its chemical composition, which depends in part on the source of the particles ("source" is called the "parent rock," thus rock parents give rise to dirt offspring). If the particles came from rocks with high iron oxide (iron combined with oxygen) content, for example, the resulting dirt may be quite red, as in much of central Oklahoma. In fact, because of the chemical makeup of the parent rocks, dirt may have oxides of several elements such as aluminum, silicon, calcium, potassium, sodium, and magnesium. Chemical makeup influences dirt's properties. Clay particles, for example, in addition to being small, have iron or aluminum oxides. Particle distributions and chemical makeup of rock and dirt are of great importance to geologists, who can use their knowledge of dirt to figure out everything from the habitat of extinct animals to the behavior of murderers and their victims.

The third important property of dirt is its organic content, that is, living organisms and their products. Soils are biological communities in the truest sense. Dirt contains a staggering diversity of organisms, and in many cases equally staggering numbers of individuals. A partial, brief, listing of soil dwellers includes: bacteria, algae, fungi and their spores, amoebas, nematodes, Earthworms, tardigrades, mites, rotifers, insects, rodents, tapeworm eggs, roots of surface vegetation, pollen, and seeds. Seeds buried, but still alive and able to germinate if given the chance, are said to be deposited in the "seed bank." Nature may draw on its deposits in the seed bank to repopulate exposed dirt (pioneer plants also travel around, as seeds, and find exposed dirt to live in).

A brief list of organic products in dirt includes bark, leaf pieces, cell wall pieces, egg shells, hair, and dead bodies. A common organic component of dirt is feces. All animals, no matter how large or small, defecate regularly, and the overwhelming bulk of their feces ends up in the dirt (except in the case water dwelling species, whose feces end up in the water). In fact, insect and worm feces are among the most frequently encountered components of our environments.

Strange as it seems, some people like to eat dirt. Periodically their dirt eating habits get reported in newspapers. It's not at all clear from these newspaper stories that purposeful dirt eaters know they're eating everything from tardigrades and rotifers to beetle and rat feces. Of course a lot of us eat dirt accidently, too, but accidental dirt is no different from purposeful dirt, at least in terms of what it contains. Tardigrades and rotifers are beautiful, mysterious, microscopic animals; they can't talk, at least in our language, but probably don't appreciate getting eaten, purposefully or otherwise. Collectively, earthworms consume truly massive amounts of dirt. In fact, earthworms eat dirt and digest the organic materials from it, which is why areas with "rich, moist, organic, soils" usually have a lot of earthworms. We don't know whether earthworms like the taste of tardigrades, rotifers, fungal spores, and rat feces; we do know that worms seem to thrive on such a diet.

Charles Darwin is a famous historical figure, mainly for his publication of the theory of evolution by natural selection in the middle 1800s. [He's also, amazingly enough, still a controversial figure for having published this theory, long since established as a scientific fact.] What most people don't know about Darwin, however, is that he also published classic volumes on barnacles and earthworms. It's the thin (153 page) earthworm book, entitled *The Formation of Vegetable Mould through the Action of Worms with Observations on their Habits* that concerns us here.

Darwin started, as he did with a number of other organisms, by keeping worms in the lab. There he studied their behavior and their mental powers. As a result of Darwin's worm studies, we now know that earthworms are prodigious consumers of organic materials and herculean dirt movers capable of burying large stones. From such study he concluded that their entire surface of England had passed through worm intestines, and that it would continue to do so forever. Then he extrapolated from England to the entire world. Of course there are astronomical numbers of worms on Earth.

Probably I should point out that Darwin was right. Earthworms are notorious for their dirt-moving, indeed dirt-processing, capacities. One of the reasonably important ecological problems in the United States is the invasion of foreign worms. In many parts of this country, native earthworms have been displaced completely by European worms, which have somewhat different habits and ecological requirements. There is no way to predict the outcome of this invasion, but the outcome is of more than passing interest to ecologists. Why? Because dirt quality and topsoil renewal are major concerns of agriculture, and the agricultural health of a nation is of major importance to everyone who buys groceries.

Earthworms are one of, if not the, key ingredient in the production of topsoil. Good dirt, i.e., dirt with lots of worms, of course, is crucial to the agricultural economies of nations. Agricultural economies tend to dictate all other facets of nations' well being. Ecologists, even ten minute ones, have little trouble equating the good life with a large supply of worm-turned dirt.

Dirt gets moved around by various factors in addition to worms, however, mainly wind, water, and larger animals. Some of these movements can be considered "local disturbances," as for example, when a gopher pushes up a mound of dirt in the middle of a golf course. Every time a tree falls over, the roots get jerked up and this event produces a large hole of exposed dirt. One of the most general characteristics of life over most of the Earth's surface is that bare dirt doesn't stay bare for very long. To say that "vegetation covers the Earth" is to proclaim rather common knowledge. But vegetation *literally* covers the Earth (okay, there are a few bare spots). Seeds and spreading plants are common enough to convert most piles of bare dirt into weed patches in a hurry. The plants that first occupy dirt piles are often those with pioneer characteristics such as small easily transported seeds, relatively low water needs, relatively few roots, and relatively large leaves in the seedling. Eventually, having been changed by the pioneers, the dirt pile begins to support more permanent kinds of plants, a process known as succession. Back to dirt.

Floods carry massive amounts of dirt, too, although these disturbances are often regional instead of just local. In 1993, for example, big floods throughout much of the Midwest deposited many inches, and in some cases several feet, of dirt and sand on what had been fertile "flood plain" farmland. Well, one of the reasons flood plains are fertile is because they get flooded, and everything from tardigrades to tree branches to rat feces gets left behind with the soil particles. Then all the organic stuff rots, the earthworms move in, and in a few years the flood plain is covered with a "rich, moist, organic" soil. Then the farmers move in, too. Periodically floods return to flood plains, of course, but when the Missouri River dumps three feet of sand on some corn field, we tend to call it a disaster instead of a renewal. To the large, short-sighted, human, the flood is a disaster; to the small, blind, earthworm whose species has been here for several million years and likes rotten fungus and rat feces, the flood is a renewal.

Large scale human activities often seem to start with digging in the dirt. Major construction (cities and highways), repairs (roads, sewers and water mains), and of course agriculture begin with a bulldozer or a plow. In this sense we are tied tightly to our planet and must alter her very essence in order to make our homes and grow our food. In a later chapter I address the question of whether we live by the same rules as beetles, and it's not likely to come as much of a surprise that the answer is "yes."

One of those rules is that like beetles, as well as gophers, termites, and a very long list of other organisms, we must move dirt in order to live, at least in our present state of civilization. All brick and stone buildings, including our great libraries, were once dirt. If you're walking through a building with marble floors, you may see fossils in that rock that used to be dirt on the ocean floor. If that building is a concert hall, and you're listening to an amazing performance by a world renowned string quartet, that violin, too, is made of wood, in turn nothing more than dirt and air assembled first by genetic information in some tree followed by a skilled instrument maker decades later. And, you can say the same thing about the cello.

5. What is water?

*Leave a log in water as long as you like: it
will never be a crocodile.*
　　　　　　　　—Guinea-Bissau proverb

Water is a chemical compound formed from two a-
toms of hydrogen and one atom of oxygen. Water is present
on Earth and elsewhere in the universe, in a number of physical forms ranging from gas (water vapor) to solid (ice).
Water is an excellent solvent; that is, many substances will
dissolve in it, so many, in fact, that water often is called "the
universal solvent." Beer, wine, and coffee with sugar are all
examples of water containing dissolved substances. Water
also is capable of carrying particles of various kinds in
suspension (as opposed to in solution). Muddy water is an
example of water carrying soil particles in suspension; if you
put muddy water in a jar, the mud in suspension will usually
settle to the bottom. Indeed, our own bodies are also prime
examples of water carrying large amounts of dissolved and
suspended substances such as proteins, fats, carbohydrates,
and nucleic acids. Water contains two hydrogen atoms and
one oxygen atom; the chemical formula is H_2O.

The familiar physical forms of water are typically
described as gas (water vapor), liquid (water), and solid (ice).
But water exists on Earth in a wide variety of physical
packages whose forms have an impact on our lives. Not surprisingly, various human languages have wonderful words
for these water forms. Such words include clouds, rain, snow,
glaciers, rivers, streams, bayous, lakes, seas, and oceans.
These water words are often preceded by modifiers: dark,
torrential, blowing and drifting, Alpine, rushing, meandering,
lazy, man-made, stormy, and open, respectively, for example.
There are so many water words and modifiers that it would
be reasonably easy for any teacher, from elementary school

to university professor, to teach ten minute ecology, as well as literature at the same time, simply by exploring our interactions with the aquatic world through these types of words and descriptive phrases. As an aside, I'm not sure we have so many colorful words for dirt, but we have quite a few for air.

Naturally occurring water falls into three general categories: fresh, brackish, and salt. Fresh water (also freshwater) is typically anything but fresh, containing, as it often does, bacteria, algae, uncountable insects, microscopic crustaceans, worms, dead toads, and duck feces. So-called fresh water does not, however, contain much salt. Salt water, i.e., as in the oceans, does contain "salt," which is actually several different kinds of salts, including sodium chloride or table salt, calcium chloride, magnesium chloride, and some of virtually every chemical element present on Earth.

Salt water also contains bacteria, algae, crustaceans, worms, dead fish, whale feces, and a doubly-astronomical number of sperm and eggs. A lot of marine animals simply release their gametes into the sea. So when you go swimming in the ocean, depending on when and where this dip takes place, you're also swimming through a whole lot of sperm and eggs. Brackish water is diluted sea water, i.e., partly salty. Brackish water occurs along the coasts where freshwater wetlands, rivers, and streams drain into the ocean and the fresh water mixes with sea water. A surprising number of plants and animals thrive in brackish water, and our coastal marshes are consequently rich, productive, and exciting places (biologically speaking).

In contrast to fresh water, sea water is remarkably constant throughout the world in terms of its chemical make-up. The ocean is relatively timeless and stable. The ocean is also very large, and so runoff from the small amount of land on Earth doesn't influence it a whole lot, except locally. Fresh water, however, occurs mainly in small amounts surrounded by relatively large pieces of land. This surrounding land varies enormously in terms of its geological history, its chemical makeup, and the way humans use it. Thus fresh water is equally varied, according to its source, because of all the different kinds of stuff that drains into it every time the rain falls.

Most humans get their drinking (fresh) water from local sources, including wells and reservoirs. For this reason, we are, or at least should be, somewhat aware of, if not concerned with, what actually drains into this water. So long as the stuff that gets in our natural water supplies is no more poisonous than duck feces, dead toads, and various microscopic wonders, then our treatment plants (and our stomachs, too) do a pretty good job of cleaning up our drinking water. But when agricultural chemicals drain off our nearby fields and contaminate wells, then we're in trouble and end up buying water in jugs filled up in presumably safe places like Arkansas.

To the average human, water's movement is perhaps its most obvious and at times most dramatic property. Floods, torrential rains, hurricanes and typhoons, thundering surf, are all examples of water in motion. For a ten minute ecologist, however, the so-called water cycle is a phenomenon that needs to be understood. Any "cycle" is a process that "goes around," that is, it occurs over and over again, with some of the same raw materials being used and same products being produced. In the case of water, what's present on Earth today is about what's always been present since the planet took on its solid, cooled, form about 4 billion years ago.

Thus the Earth as a whole doesn't make or destroy water, it simply uses water over and over again, changing the form of this water in the process. So the water that participates in the water cycle is the same water that's been participating in this cycle for hundreds of millions of years. [As an aside for my ten year ecologist friends, yes, green plants do break water molecules apart during photosynthesis, but we all know that animals recombine the parts and produce "metabolic water."]

"Water cycle" is a term that refers to the movement of water through and over the ground, through living organisms, into the air, and back again to the ground where it again becomes available for use by organisms. One of the primary players in this cycle is the sun, whose energy supplies the heat that in turn stimulates the evaporation of water. Even if there were no living organisms on Earth, water would still move between the air and land because of evaporation and subsequent precipitation, both governed by local temperature conditions. But Earth is pretty much covered with plants, which tend to lose water (a process called transpiration), mainly through pores (stomata) in their leaves. Many if not most plants have structural features, sometimes wonderfully complex, to help prevent water loss. But collectively they lose water anyway.

Animals also breathe out water, and also have various adaptations to conserve it, but like plants, animals collectively lose water too. So water moves directly from lakes, streams, ponds, and ditches into the air by way of evaporation, and water also moves into the air through the leaves of plants and lungs of animals. Eventually this water returns to the Earth and its living organisms as mist, rain, hail, sleet, and snow. When it comes back, we drink it, welcome it, mop it up, curse it, or shovel it, depending on where it falls, when it falls, in what amounts it falls, and whatever else we had planned for the time and place it falls.

Present day Earth has most of its land in the northern hemisphere and most of its water in the southern hemisphere, a fact that anyone can discern instantly simply by looking at a world map. Although this relative distribution of land and water was not always the case (see chapter 20), the continents have been in their present positions long enough to experience several cycles of hot and cold climates. The cycles we know most about occurred during the Pleistocene epoch (Ice Age), the name geologists apply to about the last three and a half million years of the Earth's history. During the Pleistocene, Earth experienced several periods of glaciation that we give names to, such as Nebraskan, Kansan, Illinoian, and Wisconsin (in Europe they're called Günz, Mindel, Riss, and Würm periods, respectively). During these periods, vast amounts of water were frozen into glaciers that flowed southward well into what is now the United States, scouring the land and leaving stone rows called moraines.

During interglacial periods, one of which we're living in now, the Earth warmed, glaciers melted and retreated, and previously ice-covered land became covered instead with vegetation. Most of this vegetation, where the glaciers used to be, is evergreen forest (see chapter 13). The take-home lesson from both the water cycle and the Pleistocene is that cycles need not occur regularly at weekly, monthly, or even annually. Cyclic events, particularly in the case of water, can occur continuously, as in the water cycle, or over several thousand year cycles, as in the glacial periods, or, for that matter, on any time scale in between.

The interaction between humans and water has historically been of major cultural importance. Polynesian islanders became adept at using the ever-present oceans as a highway and a source of food; Inuit made their houses of ice and relied heavily on blubber; ancient Egyptians dug canals; and today massive irrigation projects reveal our efforts to protect ourselves, and our economies, from the vagaries of water's natural movements. Throughout much of the western United States, streams and rivers are dammed for flood control and irrigation. But the lakes thus formed are silting in at measureable rates, obvious evidence that even the tallest mountains eventually wash down into the sea.

The ocean, of course, has served as an avenue for exploration, war, and empire-building, a source of literary inspiration, our combination grocery store and sewage plant, an arena for athletic endeavor, and a place to take a vacation.

The ocean continues to be one of the world's last frontiers, and although our technology has become quite sophisticated, we still have a lot to learn about the ocean and the organisms that inhabit it. Because we're large, short-sighted, economically-driven animals that are afraid of larger predators, we tend to view the ocean in terms of its sharks, whales, and commercial fisheries. A more proper view, however, might well focus on bacteria, algae, and diatoms. The tiniest organisms ultimately fuel the ocean food chains which in turn support the larger species from squids to whales. As in the case of dirt, our understanding of water begins at the microscopic level, an assertion that might well be true not only of dirt and water, but of virtually all ecological phenomena.

6. *What is air?*

It is a mild, mild wind, and a mild-looking sky, and the air smells now as if it blew from a faraway meadow . . .

—Herman Melville (*Moby Dick*, 1851)

Air is a mixture of gasses. Gas is technically a fluid that has no shape and has the capacity to expand indefinitely. This air fluid (gas mixture) is actually made up of molecules that are rather widely dispersed. Nearly 80% of these molecules are chemical element nitrogen. About 21% of the gas in air is oxygen. The remaining part of air includes a very long list of gasses: argon (about 1%), hydrogen, methane, helium, krypton, carbon dioxide, carbon monoxide, ozone, xenon, neon, nitrous oxide, water vapor, hydrogen sulfide, sulfur dioxide, and ammonia, to name just a few. The layers of gas that surround Earth are called, collectively, the atmosphere.

The amount of the various gasses in air is influenced by the activities of organisms that live on Earth. It is almost common knowledge that oxygen is produced by green plants and consumed by animals, whereas carbon dioxide is produced by animals and consumed by green plants. But not all organism-produced atmospheric gasses are so familiar to us as oxygen and carbon dioxide. Methane, for example, occurs in outer space and on other planets, but it is also produced on Earth by living organisms, in part through the breakdown of vegetation by bacteria. Termites are thought to contribute large amounts of methane to the atmosphere, as are mammalian herbivores (e.g. cows) through their flatulence.

Other gasses that occur in air, in part because of living organism activity, include thousands of kinds of molecules somebody or something perceives as an odor. Everything that smells is a gas, made of molecules that stimulate nerves in a particular way. The odors of perfumes, urine, feces, rotting garbage, petroleum products, flowers, the rich smells of marshes, the piney-woods smell of northern evergreen forests, the characteristic odor of mildew, the aroma of yeast in a half-cup of warm water, the unmistakable blast of a dead skunk on the highway, are all produced by gasses that make up air. In addition to perfume and sewage, there are many molecules that humans cannot smell but other organisms can, although "smell" may not be exactly the right word. Pheromones, for example, are molecules, produced by animals, that influence the behavior of other animals, typically members of the opposite sex. Although people may not be able to smell the pheromones produced by a female moth, the male moth can sure detect these molecules in air and usually flies toward their source.

Like any fluid, air can be either hot or cold. Cold air is heavier than warm air, and when the two come into contact with one another, the mixing sometimes gets turbulent. When air gets warm, it expands, rises, or pushes on the cooler air around it. When air gets cold, it condenses, falls, spreads out, and pushes on the warm air around it. The air around the equator is, on the average, warmer, thus always lighter, than air in the more temperate zones. The air at the poles is, of course, always relatively cold, thus heavy. Heavy air pushes down on Earth more forcefully than light air, therefore generating more pressure on us. Thus we use the terms *equatorial low* (pressure) and *polar high* (pressure) to describe these different air regions.

On a large scale, heavy cold polar air flows toward the equator, gets warmed, rises, and then gets pushed back toward the poles, all in a grand cycle. Earth's atmosphere is in constant motion partly because of this warming and cooling of air, but also because of the planet's rotation. This rotation disturbs the large scale air movements and causes wind to blow predominantly from the northeast in the northern hemisphere and from the southeast in the southern hemisphere. If you've watched old pirate or sailing ship movies, or read the exciting books on which these movies are based, you have heard the term *trade winds*.

Trade winds are the predominantly northeast or southeast winds produced by polar highs pushing themselves down into equatorial lows and at the same time being swirled by the Earth's rotation. The winds are called trade winds because they pushed and pulled sailors along their trade routes with predictable regularity. These winds also tend to disappear near the equator. When the trade winds stopped, sailing ships did too, usually getting becalmed near the equator, far from land, food, and drinking water. Early sailors called these places the doldrums.

We now use the term doldrums as a metaphor to describe a feeling of being stuck and feeling down and useless. Samuel Taylor Coleridge drew upon such an involuntary becalming for his famous poem *The Rime of the Ancient Mariner*, which is the source of another metaphor, namely that of the albatross [around one's neck]. As you can see from this discussion, ecologists, even ten minute ones, are not above noting how human language is often enriched by our interactions with the natural world, including air.

Anyone who watches weather reports on television sees areas of high and low pressure spread across the continents and changing positions. In high pressure areas the air is heavy; in low pressure areas the air is light. We don't really understand all of the underlying reasons why air is heavy or light at a particular place, but temperature obviously has something to do with it. High pressure can be thought of as a pile of air which eventually collapses under its own weight and spreads out. Low pressure can be thought of as a sink into which surrounding air drains.

Low pressure often portends interesting weather because both warm and cool air can be sucked into a low pressure area. Such mixing produces storms, which are bad for people who are afraid of lightning or getting hit by tornados, but good for people like artists who like dramatic cloud formations. Blown-down trees and lightning-started fires are also natural phenomena that create disturbances. Such disturbances in turn generate temporary environments that some specialized plants require. Thus in an indirect way, for example, prior to the settlement of North America by Europeans, low pressure areas had a direct influence on the native plant communities through lighting-started fires, which were ever-present although not always frequent.

In general, winds rotate clockwise in a high pressure area in the northern hemisphere, but counterclockwise in a high pressure area in the southern hemisphere. Exactly the opposite happens in low pressure areas. These swirlings of high and low pressure are felt by all living organisms—plants, birds, insects, spiders, and bacteria, as well as humans. Many plants have seeds adapted to wind dispersal, the common dandelion being a prime example. Spiders are also notoriously successful wind-travelers; newly hatched spiders spin out long strands of web that are caught by the wind and carried great distances, usually with the baby spider still attached. For this reason, spiders tend to be among the first colonizers of areas stripped bare by fires, floods, and volcanic explosions. Of course spiders are also carnivores, so unless some food insects also arrive, wind-sailing spiders tend to starve. But flying insects tend to get blown around, too, especially strong fliers such as dragonflies, and light but good fliers such as mosquitoes and houseflies. So the atmosphere can be thought of as an avenue of dispersal for many kinds of organisms.

Air also contains solid particles: dust, smoke, pollen, fungal spores, pieces of spider web, tiny flying insects, and bacteria, to name just a few. "Dust" is a catchall word, of course, that includes as long a list of items as found in dirt. Dust is particularly interesting because of its colors (partly responsible for beautiful sunsets and sunrises), what it can tell us about events happening far away, that have already happened far away in time, or are likely to happen in the future. Scientists studying dust can determine whether a nuclear explosion has occurred, or a volcanic eruption, or erosion in an arid land somewhere. They can also tell us the kinds and weights of particles produced by various industries, warn us of particularly bad smog or allergy conditions, and predict the effects of wind on exposed ground. Homework for a ten minute ecologist probably ought to include study of the famous Dust Bowl photographs from the 1930s; few sources of information can so quickly teach us about the power of air to move dirt and whatever else dirt contains as those pictures.

Before closing this chapter, I probably ought to mention both the ozone layer and greenhouse gasses. Ozone is a molecule made from three oxygen atoms; there is a layer of ozone, up in the atmosphere about 12 miles high above the ground. Along with all the other stuff in air, e.g., dust and clouds, this ozone absorbs ultraviolet radiation from the sun. Ultraviolet radiation is generally lethal to living organisms and is one of the major contributors to skin cancer in humans, so this ozone layer is reasonably important to life on Earth. Certain industrial chemicals released on the ground move upward into the atmosphere where they interact with ozone and destroy it, in effect destroying the shield that protects life from ultraviolet radiation. For this reason, most Americans sometimes can read as much about ozone in their daily newspapers as they can find in freshman biology texts.

Greenhouse gasses are mainly carbon dioxide, which virtually every animal, including humans, breathes out regularly. Stove gas (methane), which is also produced by termites, cow flatulence, and swamps, is another greenhouse gas, as are water vapor and various other industrial chemicals. Collectively these gasses are called "greenhouse" because they absorb heat and radiate it back to Earth, thus creating an effect that operates like a global greenhouse. Under "natural" conditions, the amount of carbon dioxide in the atmosphere varies according to season in the northern hemisphere where most of the land, and thus most of the plants, are located.

Plants use carbon dioxide, so the amount of this gas in air tends to be lower in summer than in winter. But clearing of forests and burning of coal and petroleum both increase the amounts of carbon dioxide in the atmosphere, thus increasing the amount of heat trapped and radiated back to Earth. In a doomsday scenario, this increased heat melts the polar ice caps and Los Angeles is flooded. But if our knowledge of geology is accurate, Los Angeles will probably be flooded or pushed under some crustal plate anyway at some time in the future. Whatever the effects of human activities on global climate, the insurance companies who end up paying for hurricane and flood damage are, at this very moment, worrying mightily about these effects.

Finally, air is a relatively hostile environment when viewed in terms of body fluids. Virtually all organisms tend to lose moisture in air, and during evolutionary history, colonization of land was always accompanied by acquisition of some features that helped preserve body water. The hardened skin of reptiles, the waxy hard exoskeleton of insects, and the human kidney are all examples of such water-conserving features.

Dirt, water, and air—three general categories of non-living components that make up our environment are all seen to be quite complex and variable entities, typically containing a lot of organic materials. Their qualities often dictate our relationships with them, and in this sense all are similar in their influences on our lives. We also have the power to change those dirt, water, and air qualities, and certainly exercise this power at least on a local scale. But we're all bathed constantly in air which is always moving on a hemispherical, if not global scale. For this reason, air symbolizes perhaps better than any other factor our shared dependence on planetary properties that may seem so distant from our immediate daily concerns.

7. Who eats whom?

> Edible, adj. *good to eat, and wholesome to digest, as a worm to a toad, a toad to a snake, a snake to a pig, a pig to a man, and a man to a worm.*
>
> —Ambrose Bierce (from *The Devil's Dictionary*, 1906)

A feeding relationship, in which one organism eats another, is called a "trophic relationship." Thus a study of trophic relationships is a study of who's eating whom, under what circumstances, and in what amounts. The variety of feeding mechanisms and devices, among plants, fungi, animals, and microbes, is simply astounding, but basically, organisms obtain two things from their environments: energy and molecules. For most organisms, the energy is in the form of molecules ("Eat some sugar! Get some energy!"). Although green plants get their energy directly from the sun, they use that energy to build molecules that contain energy.

The concept of energy being contained in molecules is one of the most central and important ones in all of biology. Organisms need energy in order to build and maintain their bodies, just as you require energy (e.g. electricity) to build and maintain your house. Some molecules, in addition to containing energy, also serve as sources of building materials. What gets built from these materials? Cats get energy from mice (they eat) then build more cat, for example. Cats also get energy from cat food bought at the store, of course, and use that cat food to build more cat, so when you buy cat food, you're actually buying artificial mice.

In matters of eating one another, there are three categories of organisms: (1) producers, (2) consumers, and (3) decomposers. All the organisms we commonly call green plants are producers. Producers don't produce energy; they produce and store only an accessible *form* of energy. That *form* is a green plant. Forms of energy are vitally important to living organisms. If you wanted to toast a piece of bread and the only energy in your possession was in the form of sugar, you'd be out of luck. Similarly, sunlight, a form of energy that gives you cancer, cannot be used directly to satisfy your personal energy needs. Plants take the sunlight and convert it to energy, which you can then use. Conversely, plants [usually] can't use electricity and gasoline, *forms* of energy that most of us use every day in reasonably large quantities because of our technological achievements.

All of the organisms we commonly call animals are consumers. Consumers eat organic materials, e.g., plants, other consumers, or parts of them. So in general, consumers eat producers. This feeding relationship creates a "link" between consumers and producers; not surprisingly, a series of such links is called a chain, specifically a food chain. Seeds eaten by mice which are in turn eaten by owls constitute a food chain. A dead owl, however, is typically first consumed by insects, which then become another link in the food chain. But decomposers (bacteria and fungi) are pretty democratic; they'll eventually decompose both producers and consumers. Out in nature, all this eating and decomposing produces food chains that link many plants and animals together. Also not surprisingly, an interconnected mess of food chains is called a food web, which is probably a more accurate metaphor, than a chain, to describe the trophic relationships between organisms in a natural environment.

Ecologists view green plants in terms of the molecules they contain (as well as in other terms, described in other chapters). Thus the green plant can also be called a package of carbohydrates, lipids, amino acids, and vitamins. Technically, these molecules are where the plant's energy is stored. The fact that you can burn a log illustrates the fact that a green plant is a form of (i.e., contains) energy. The burning releases the energy. If you used the burning log to cook a hamburger, then some of that log's energy would have been used to alter the condition (cook) of the once raw hamburger.

Without green plants there would, of course, be no animals, at least of the familiar types, and life, if it existed on Earth, would be quite different than we know it to be today. However, because of the various properties of star (sun) light, it's worth speculating about whether life on other planets would include green somethings. Why is this speculation a legitimate one? Because the molecules that trap sun (star) light make the trappers look green. Thus we expect that in a far off galaxy, producers might well be as green as they are in your local corn field or vacant lot.

Decomposers are the fungi and bacteria that break down dead producers and consumers into smaller and smaller, indeed back into molecule-sized, particles. Decomposers typically secrete digestive enzymes that function to break up the food into molecule-sized bites ("spit" on their food); then the decomposers absorb the digested materials ("suck" it back up). The decomposers use these absorbed molecules to make more decomposers (bacteria reproduce by dividing; fungi grow and produce spores).

Toadstools and mushrooms are the spore-producing parts of some fungi, so every time you see a toadstool appear on your lawn, you're actually seeing part of an underground decomposer at work. And if you've had a tree removed from your yard, the ring of toadstools that appears regularly on the same spot is actually evidence that decomposers are at work, beneath the ground, getting rid of the stump and roots. Decomposing bacteria and fungi are extraordinarily common and widely distributed, and so all organisms, even bacteria and fungi themselves, end up as decomposer food. Without them, we would be living in a very strange world.

Consumers that consume green plants are usually called herbivores or primary consumers. They're herbivores because they're eating "herbs," a term that refers mainly to non-woody vegetation, although ecologists refer to animals that eat woody plants—e.g. trees and shrubs—as herbivores, too. Herbivores also are called primary consumers because they're consuming producers. In other words, they're eating at the primary, or first, or metaphorically speaking, the lowest, level [of the food chain] at which the sun's energy is made available to other life forms on Earth.

Consumers that eat herbivores are called carnivores ("meat" eaters), or secondary consumers, because, metaphorically speaking, they're living at the second, or next highest level at which the sun's energy is made available to us. That is, secondary consumers eat primary consumers that in turn eat producers. It's important to remember that energy in primary consumers was energy originally stored in green plants, which in fact is part of the energy that fell on Earth as sun (star) light.

It is a general rule that in nature, the mass of producers is greater than the mass of primary consumers, and the mass of primary consumers is greater than that of the secondary consumers. We usually think of mass, also called "biomass," as weight. In many cases, the number of producers is also greater than the number of primary consumers, and the number of primary consumers is greater than the number of secondary consumers.

This decrease in mass and/or numbers of species as one moves up the food chain from producers to primary consumers to secondary consumers is often drawn as a pyramid made of stacked layers. The producer layer is widest, the primary consumer layer is less wide, and the secondary consumer layer is less wide still. You could obtain the same type of image by stacking a quarter (primary consumers) on a silver dollar (producers) and a penny (secondary consumers) on the quarter. Then you'd have a small pyramid. Logically, ecologists call this mass-to-number relationship between producers and various levels of consumers a "food pyramid."

Food pyramids are conceptual and theoretical structures, just like food chains and webs. Their properties are observed in nature, however, in much more practical terms. For example, the grain necessary to build and maintain a mouse weighs more than the mouse; the mice necessary to build and maintain an owl weigh more, collectively, than the owl. But in order to discover this relationship between biomass and trophic level, we have to count all the mice that the owl eats in a lifetime. The grass necessary to build and maintain a zebra weighs more than the zebra; the zebras required to build and maintain a lion weigh more, collectively, than the lion.

Anyone could discover this biomass-trophic level relationship simply by buying a kitten, weighing it regularly, and weighing the amount of cat food purchased to keep it alive throughout its lifetime. (Don't forget to estimate, and add in, the amount of energy the mother cat used to build the kitten before you got it.) The weight of this cat food could easily be up to ten times the weight of the cat. There are also more seeds on Earth than there are mice; there are more mice on Earth than there are owls. There are more grass plants on Earth than there are zebras; there are more zebras than there are lions. There probably are more cans of cat food on Earth than there are kittens, too.

Why do these mass and number relationships generally hold? Because most energy is "lost" when it is transformed from seeds into mouse, from mouse into owl, or from zebra into lion. Where does the lost energy go? In simple terms (some would say even simplistic terms!) it goes to keep the mouse, owl, zebra, and lion warm. That is, much of it is lost as heat. This loss of usable energy during transformation is a fundamental law of the universe, and we humans obey it as surely as seeds, mice, and owls. Thus coal can be transformed into electricity (coal-fired power plants) which can then be transformed into heat (electric furnace in your house). The coal was originally, many millions of years ago, plants, so it can be considered fossil sunlight energy. During the transformation of coal to electricity, some, indeed most, of the fossil sunlight energy is lost to us, and more is lost during the conversion of coal into electricity.

Some animals are called "top predators" because they eat other animals that in turn are said to live "high on the food pyramid." Often "top predators" are large, dramatic, sometimes exotic, animals that are often shown on television: lions, cheetahs, and eagles are good examples. So are great blue herons, a common animal throughout much of North America. I've often thought that the great blue heron must be near the top of the food pyramid because it eats frogs. Imagine, for example, that this frog eats a dragonfly that's eaten a mosquito that's sucked a human's blood made from beefsteak which is in turn made from grass whose energy came originally from sunlight.

In this example of a food chain we have grass (1), cow (2), human (3), mosquito (4), dragonfly (5), frog (6), heron (7); the numbers in parenthesis represent the cumulative energy transformations necessary before the world sees great blue herons. I've always been surprised that there are so many herons in the world because they live, energetically speaking, so far away from the sun, even though they may nest in the tops of trees, with their chicks baking in sunlight they can't use. But then herons eat a lot of small fish, too, and the world's rivers and ponds are generally filled with small fish.

When assessing who eats whom, one needs to consider where, in the world, the assessing is being done. In other words, are we considering the tropical forests, the polar oceans, the grasslands, or some other type of habitat? One also needs to consider the scale at which we assess who eats whom. For example, in a roadside ditch, all of the producer, consumer, and decomposer relationships may be present, but at a microscopic scale. Conversely, out on the African veldt, seen so commonly on television nature shows, the producers and consumers may be quite large (but the decomposers are still generally pretty small).

And out in the ocean, the food pyramid is mostly hidden from us. On the surface (pun intended) oceans seem to operate differently from terrestrial environments, but the seeming difference is due to the fact that most of us are not very familiar with oceanic producers and consumers. Much of the ocean production is carried out by diatoms, mostly microscopic, photosynthesizing, one-celled organisms that build beautiful silicon protective shells. Shrimp-like crustaceans, known collectively as krill, feed on diatoms, and a large variety of consumers, ranging from whales to seabirds to squid, feed on krill.

All of this eating and decomposing results in the "movement" of energy and molecules from one organism (for example, a mouse) into another (a cat). Ecologists call such energy movement "flow." Thus energy flows through an ecosystem, following the paths that are established by a food web. It's the molecules themselves, some of which are used to build plants, animals, and microbes, that actually do the flowing, and they carry the energy along with them. A cursory glance at the daily newspaper usually reveals at least one story about energy supplies. Although the "energy" in this case is usually electricity, oil, gasoline, or coal, all of it except electricity produced by nuclear power was once sunlight trapped by green plants. Even hydroelectric power is ultimately dependent on the sun and atmosphere. For this reason, energy issues will likely remain in our newspapers, and in our political arenas, for a very long time.

8. Who beats whom?

Show me a good and gracious loser and I'll show you a failure.
—Knute Rockne

Charles Darwin believed that competition was the driving force of evolution. Post-Darwinian scholars and politicians alike picked up on this idea, converting it, at times, into theories such as social Darwinism that today we find morally lacking. However, competition, especially on the athletic field, is an ingrained part of our society and if the historical and archeological records are correct, this has probably been the case for several thousand years. But for ten minute ecologists, competition needs to be defined in ways that go beyond athletics, war, and politics. We need to ask, and answer, questions such as: For what do organisms compete? How is this competition manifested? What are the outcomes of competition? And so forth. We also need to ask a more fundamental question: Do organisms always compete with one another? This last question is easy to answer: no (or at least probably not). And if, in a particular case, the answer to this last question is "no," then for that case the rest of the questions become irrelevant.

In any discussion of competition we need to agree upon the number of species involved. Competition between members of a single species is called intraspecific (= within a species; among members of a single species) competition. Competition between members of different species is called interspecific competition. The average person who has not studied much biology typically thinks of interspecific competition when the subject is raised. After all, on the athletic field Bears compete with Lions, Eagles with Falcons, etc. In nature, however, intraspecific competition (e.g., between different lions, or between different bears) is most often the situation, simply because organisms of a single species are most likely to require the same resources. They're also likely to seek mates within the same population. So here is a general rule about competition: *we compete most strongly with those most like us.* This same general rule applies to all organisms, although not necessarily to competition for all resources.

There are two conditions that must hold before organisms can be considered to compete in the typical sense: (1) these organisms must use a common resource, and (2) that resource must be limited in some way. These two conditions are essential ones for "natural selection" as defined by Charles Darwin. Another essential condition is that members of a species vary in their ability to compete for the resource. When these three conditions hold, then "nature" is considered to "select" those variants that are most capable of winning in a competition for a resource. But "winning" does not always mean what the average person thinks it means. No, in this case, "winning" means reproducing more than your competitors. The score is kept in numbers of surviving offspring. And, to survive, in this sense, is to reproduce in turn.

Not all competition is between members of one single species, however. Sometimes several species compete for a limited resource. A good example of such a resource might be all the flowers on a tropical plant. The organisms competing for nectar from these flowers might include hummingbirds as well as several kinds of insects. Together, the species utilizing a common resource are called a "guild." The guild of species that rely on tropical flowers might well include hummingbirds, bats, and bees.

Historically, the term "guild" also applied to a group of people with similar interests and ways of earning a living, e.g. the medieval group of craftsmen such as shoemakers and brewers. From that usage, we have "guild hall" (union hall) and "guildsman" (member of an association of similar craftsman). So it's not very difficult to see how the term might be extended to apply to a group of species that "work" on a common resource and assemble there. Nor is it difficult to imagine these guild species competing, much as medieval shoemakers must have competed for business in a particular town.

The average citizen, perhaps one who watches many nature shows on television, sometimes forms an impression that natural systems are always driven by competition. A ten minute ecologist, however, should remember that not all resources are necessarily limited. Grass on a grassland, water in a tropical rainforest, and, some believe, food available to tapeworms in a dog's intestine, are examples of resources that are not necessarily limited. So one might ask: Did bison evolve their particular characteristics as a result of competition for grass on the Great Plains? The answer is probably no. Our best guess is that bison competed mostly for mates instead of grass blades and developed certain physiological and behavioral traits in response to climate conditions. They probably did not become such powerful bulls and skillful mothers by shoving one another aside to get at grass blades on the Great Plains where, prior to settlement, there was a whole lot of grass.

Competition for a limited resource may be quite subtle, and the long term results of such competition are not always readily apparent. Plants may compete for sunlight even though sunlight is hardly limited. What is limited, however, is the space and often the water needed for plants to grow enough to get their leaves exposed to sunlight. When you "thin out" radish seedlings from a garden row, you're reducing intraspecific competition.

Competition can be direct—a hummingbird feeding from a flower, thus preventing another hummingbird of the same species from getting the nectar from that same flower. Competition also can be indirect—a praying mantis eating a fly that then is not available for another praying mantis at another place or later time. But in both cases, one organism is doing something that influences another's life, if not actually, then potentially. Sorting through all these forms, effects, and potentialities is often a quite difficult problem, usually involving uncooperative, and often not particularly tractable, plants and animals.

Because of the great variety of organisms on Earth, the many resources for which they could compete, and the subtlety and indirectness of some interactions, the effects of competition on both the daily life of individuals and the evolutionary life of species are not always easy to discover. So it's not surprising that ecologists have spent an enormous amount of time and energy studying competition. In general, as a result of all that study, here's what they've decided:

(1) Competition is an ever-present and powerful influence on organisms' daily lives, as well as the lives of the species to which they belong. The competition can be for virtually anything an organism needs in order to survive, so long as the "anything" is in limited supply. But the outcomes of any natural competition are ultimately measured, and in fact summarized, in terms of offspring that in turn reproduce.

(2) Organisms compete with all the "weapons" available to them. Such weapons include some familiar ones such as antlers, horns, and fangs, as well as some less familiar, or at least less "weapon-like," ones such as odor-causing chemicals, colors, behaviors, and responses to environmental conditions.

(3) Competition between species may be mediated by the environment. In one of the most well known examples, two species of barnacles competed for space on intertidal rocks, with the larger, faster-growing species usually winning. But on the higher rocks, the smaller, slower-growing species won because it was more resistant to the drying effects of exposure.

(4) Competition is still considered one of the causes of evolutionary diversification, although certainly not the only cause, and (according to some) perhaps not even the primary cause. Instead, "isolating mechanisms" such as chromosome differences that inhibit fertility are considered important factors in producing diversity among plants and animals. And, of course, "mutations," spontaneous changes in the genetic material whether it be DNA or chromosomes, are essential to any evolutionary change.

(5) Plants compete for water, soil minerals, and sunlight. Plants also compete with one another by releasing chemicals into the soil, such chemicals then inhibiting the growth of other plants. A whole lot of the competition among animals is for mates.

(6) When you see some predator attacking prey, e.g. a cheetah chasing an antelope on television, the predator and prey are not competing with one another. If the cheetah is competing in this instance, the competition is with other cheetahs and the fastest one is most likely to catch the antelope. Conversely, the antelope is competing not with the cheetah, but with other antelope, again in terms of speed, in this case of escape.

In summary, organisms do compete for a great many resources, but neither the form nor the results of this competition are always obvious to the average person. A ten minute ecologist can easily train his or her eye to look for signs of competition. The best place to start looking is often in the back yard garden, and the first thing to look for is plants of the same kind placed too close to one another.

9. What is an ecosystem?

. . . nature has all kinds of models.
—Buckminster Fuller (*In the Outlaw Area* by Calvin Tomkins, *The New Yorker*, 1966)

An ecosystem is a combination of all the biological and physical properties of the natural world, usually in a recognizable area. If that definition sounds rather dauntingly academic, it's because of two reasons: first, it is an academic definition, and second, the term "ecosystem" is not particularly easy to define in a satisfactory way.

Perhaps the best way to think of an ecosystem is to envision all of the biological and physical events, e.g., plant growth, rain, temperature fluctuations, predation, parasitism, death, etc., all occurring in a relatively large geographic area tied together by some dominating physical feature. Thus one could envision, for example, "the prairie ecosystem," which would encompass a large, more or less self-sustaining, relatively flat and dry region in which perennial grasses are the dominant vegetation, characteristic native vertebrate animals are predominantly herbivores such as bison and rodents, and the major shaping physical forces are fire, wind, and extreme temperature fluctuations. If you're bored by prairie ecosystems, then of course you could envision a coral reef ecosystem, a desert ecosystem, or the most complicated ecosystem of all, the tropical forest ecosystem.

In the absence of human disturbance ecosystems tend to remain stable for relatively long periods, e.g., thousands of years. During most of Earth's history, destruction of ecosystems resulted from global events, typically of geological origin, such as the drifting of continents and the rising and falling of ocean levels. The fossil record shows us that some, if not all, continents have experienced dramatic changes over their several hundred million year histories. For example, much of America's prairie ecosystem "bread basket," the Great Plains where most of our corn and wheat are grown, used to be a large sea. For this reason, enormous aquatic reptile fossils are found in places like Nebraska. In general, such changes occur slowly, or at least at a rate most humans would consider slow. It took well over a hundred million years for Kansas to change from a vast sea of salt water populated by swimming monsters (mosasaurs and plesiosaurs) to a vast sea of grass populated by walking and running monsters (mammoths and bison).

The ecosystem concept is a highly instructional tool with which humans may easily learn a great deal about their planet and significantly enrich their trips to museums and zoos. The concept is so useful in this regard because it allows us to organize an enormous amount of information into manageable chunks. Much of this information consists of information a well informed citizen already knows, but simply hasn't thought about in an organized, ecological, way.

For example, the word "Africa" is a fairly familiar one to us, and many African animals also are quite familiar because we see them in a variety of places—zoos, museums, television, and National Geographic. In fact, we see some African animals, especially lions, leopards, cheetahs, elephants, zebras, rhinos, giraffes, and a dozen species of antelopes, so often in pictures that they tend to be a part of our stereotypical vision of the Dark Continent. The African continent, however, has many ecosystems, for example the Sahara Desert, the east African savannah, the steamy forests of the Congo River basin. Each tends to be characterized by its own mixture of temperature and moisture, as well as its own "food web" consisting of the species that live in those ecosystems and feed on one another.

In another chapter (13, "How is real estate really divided up?"), I discuss the concept of a biome, which is a relatively large area characterized by reasonably stable vegetation types. A particular biome can occur anywhere, so long as the moisture, temperature, and soil conditions that produce it also occur. The term "ecosystem" includes and infers many of the same concepts as "biome," particularly stability and relative uniformity. But ecologists tend to use "ecosystem" in a more flexible way than they use "biome," and in fact the latter term is evolving somewhat out of vogue because we now know that islands (or pockets?) of certain vegetation types can occur within larger matrices. Such occurrence is dictated primarily by temperature, moisture, and soil type. In plain words, local conditions can alter the regional vegetation patterns (e.g., golf courses in Arizona). Usually those local conditions have something to do with water, however, such as an oasis in the desert, or your living room (with your watering pail and an exotic plant).

A ten minute ecologist needs to be aware of two truly important concepts that apply to ecosystems. One of these concepts is that of *flow*; the other is that of the *cycle*. Perhaps the best analogy for flow and cycling in ecosystems is the movement of money and goods through a society. Indeed, this analogy is so useful that professional ecologists share many ideas, methods, and metaphors with the economists. Every time someone buys something, money flows and is transformed in the process. No matter whether the transaction is as simple as a child buying a piece of candy or as complex as a large real estate development, money moves ("flows") and is transformed (from cash or credit into happiness and cavities [child buying candy], to stress, livelihoods, construction accidents, insurance, and environmental hassles [the real estate development]). So commerce is an excellent model for energy flow (or vice versa).

Societies also recycle items. Rumors and bad ideas are items that seem to move around fairly regularly throughout most societies, although in the case of these purely human constructions, movement is often so disorganized that it's difficult to tell where the "cycle" is. An obvious, and perhaps more serious, example of a cycle is the movement of the element aluminum. Most large cities have aluminum recycling sites, and if these sites are not present, gleaners often dig through garbage to find cans worth money in the recycling market. Aluminum comes from the ground (it's mined), is made into various things like cans and toys, then returns to either the ground (landfill) or back into the loop of useful, bought, used, then tossed (or recycled) products. Aluminum is also a good example to use in this context because it's a chemical element and chemical elements tend to move in cycles through natural ecosystems.

Energy, carbon, and nitrogen seem to be a rather abstract concepts or items, however, compared to money and aluminum. Think of it this way: people eat in order to get the energy necessary to work and play. Well, so do pet cats, rats, birds, and worms. Cats eat rats and birds eat worms, thus obtaining the energy to do the cat and bird equivalents of work and play. Rats, of course, eat almost anything, and worms eat dirt, thereby obtaining energy to do the rat and worm equivalents of work and play. When a cat eats a rat, however, energy flows from rat to cat, i.e. into cat. Some of the molecules in the rat—for example those containing the chemical elements carbon, oxygen, and nitrogen —also move into cat.

When the cat defecates into the yard, or dies and is buried (with ceremony) in your garden, some of these chemicals, as well as some energy, get returned to the dirt. When the worm eats the dirt, the chemical elements and energy flow into the worm. When the bird eats the worm, some energy, nitrogen, carbon, and oxygen move into the bird. If the cat eats the bird, then the cycle is complete. The most important point for a ten minute ecologist to remember, however, is that in all this eating, defecating, dying, and flowing (moving), the energy is gradually lost from the system whereas the chemical elements are recycled.

The energy that is lost is replaced by the sun. That is, the sun's energy is captured by the ecosystem's green plants, then stored as plant tissue (leaves, fruit, bark, stems, roots), although some of this energy is lost in the storage process. But once the sun's energy is captured, it eventually becomes available to all those species that eat plants, and thus to those who eat plant eaters. Nevertheless, with each transformation, such as seeds into mice, some energy is lost (see chapter 7). Where does the lost energy go? It goes into heat. An ecosystem can be thought of as all these life processes, involving the movement and exchange of energy and materials as described in other chapters of this book, all operating, linked together, and constrained by conditions of climate and geography. This integrated picture of life is what an ecologist understands when he or she hears words like "prairie," "desert," "coral reef," or "rainforest."

10. Why are the tropics so complicated?

It is only against the panorama of modern research that the full value of Walter's butterflies becomes apparent.

—Miriam Rothschild (*Dear Lord Rothschild*, 1983)

The tropics are complicated primarily because they have a large supply of both heat and moisture. Maybe, in order to be scientifically correct, I should turn that statement around: areas that have relatively large supplies of both heat and moisture tend to be relatively complex environments. Thus when I use the term "tropics" in this case, I'm referring mainly to the tropical forests, the steamy, exotic, jungles of both fact and fiction. From an ecologist's point of view, these types of habitats are clearly the most complex ones on Earth. What makes them complex? A major contributor to this complexity is the multitude of different physical spaces provided by the luxuriant plant growth, the vast number of species found in these spaces, and the many different kinds of relationships between the organisms that live in Earth's warm regions between the Tropic of Cancer and the Tropic of Capricorn (23° 27' north and south latitude, respectively). The great range in elevation—from coastal wetlands to the highest Andes—also produces varying conditions of temperature and moisture that in turn contribute to diversity of plants and animals.

In areas with large supplies of both heat and moisture plants grow readily, a fact that's pretty obvious to anyone who's ever tried to grow a plant whether it be a garden beanstalk or a philodendron cutting in the house. In the tropics, however, plant growth is so extensive, and there are so many different species, that the space over the land is very complex. By complex, I mean there is great variety in the shapes of organisms and their parts. A typical mature tree in the United States, for example, has hundreds if not thousands of branches, all of different sizes, shapes and positions relative to the ground, and each branch has many leaves—hundreds or thousands in the case of some pines, spruce, and their relatives. The branches can be living or dead; dead branches can be broken off, exposing a jagged tip or sometimes a hole; part of the tree can be rotten, infected with a fungus, or attacked by insects, and each of these events adds to the difference between branches and between leaves, i.e., creates heterogeneity = complexity. The bark may be thick, providing many cracks and crevices for tiny insects and spiders. A woodpecker may have drilled a hole, lichens may be covering one side, or a fire may have scarred the trunk. And the roots, which are also quite varied, usually occupy more space beneath the ground than the stems, branches, and leaves do above ground.

Anyone can discover this physical complexity just by closely studying the nearest tree, even the one in his or her own back yard. This simple bit of exploration, for which you pay nothing and will not be tested over, is improved greatly by use of a magnifying glass. The trick to becoming a ten minute ecologist is to learn to see this local and familiar plant as a heterogeneous mixture of spaces that can be occupied by small animals, fungi, bacteria, viruses, and other plants. And, of course, the smaller the scale of vision, the more heterogeneous the world appears (thus the magnifying glass).

Hetero- is a word stem borrowed from Greek and it means "different" (e.g., *hetero*sexual—needs no explanation). *Heterogeneous* is an adjective that means "characterized by difference (i.e., diverse);" *heterogeneity* is a noun that means "the condition of being diverse." That single tree in your back yard, by its very presence, with its various branches, leaves, etc., produces heterogeneity—in your garden, in the space defined by your property lines, and in your life. So when we look at them closely, large plants even in temperate zones are seen to be quite complicated structures. This complexity is multiplied many times over in the tropics because there are so many plants, and of so many different kinds. All this diversity and growth produces an astronomical number of ecological niches.

In ecological parlance, the word "niche" means a defined set of conditions under which a species exists (see chapter 2). In a heterogeneous environment, there are many niches, thus many "places" for various species to live. Over evolutionary time (see chapter 15), small organisms adapt to these niches, or, in a more technical and abstract sense, create niches by living in them. Each organism's unique ecological niche is an analogy to an office, and a life history can be thought of as a kind of natural career. So the tropics, with their physical and botanical heterogeneity, provide jobs for vast numbers of tiny species, mainly insects, mites, spiders, and especially ants. Those "jobs" are the roles these organisms play in the ecosystem and have names like "producer," "consumer," "predator," "symbiont," and "parasite." When you watch a nature program on television, you're actually watching organisms trying to make a living at their particular roles, using mostly only the genetic information they inherited (the vertebrates may also be using information they learned on the job).

It's a general rule among organisms that each species has its own typical, and sometimes very specific, life history, in addition to its own unique structure. A frog that carries its tadpoles on its back, for instance, has a different life history from one that lays its eggs in the water then hops off, never to mess with the offspring again. So with all those tropical species, we also observe equally diverse life histories. A species' life history helps define its relationships with its environment, too. For example, the tadpoles embedded in the parent frog's back skin have a different relationship with their environment than do those free in a puddle or in the water contained in a plant. So in the tropics we have not only physical heterogeneity produced by luxuriant plant growth, but biological heterogeneity produced by the many species, their many different life histories and relationships with their environments.

But life history differences are certainly characteristic of plants as well as animals. For example, in the tropical forests we find both rapidly growing trees such as balsa, and slow growing hardwoods such as teak and mahogany that are so prized for furniture, art, and utensils. We also find many plant species that have no roots, but live as epiphytes (*epi* = upon; *phyte* = plant) on other plants. Interactions between plants and animals also contribute significantly to diversity in the tropics. Fruit-eating birds, bats, and monkeys distribute the seeds that pass undigested through their digestive systems, beetles bury monkey dung containing seeds, thus increasing the chances of the seeds surviving a year, flowers and nectar-feeding pollinators co-exist in a mutually beneficial, although mutually dependent, relationship, and ants evidently harvest a significant fraction of the wild seed crop. Some flowers are pollinated only by certain insects, bats, or hummingbirds; indeed, very dramatic evolutionary histories are often revealed by flower—pollinator relationships. And as might be expected in the forest, some plants live only in clearings created when large trees fall.

The number of plant species also can be exceedingly high; in one study in Panama, for example, scientists actually marked 238,000 trees in an area 1 km by 0.5 km (about 6/10 of a mile long and 3/10 of a mile wide, a little over a hundred acres) and studied these trees for 12 years. The 300 tree species in that small area seems like a lot of diversity, but a similarly sized study area in Malaysia contained at least 800 species. A 1996 news report claimed that the record for concentrated diversity is in the Brazilian rain forest where ecologists discovered 476 tree species in a 2.5 acre plot. American temperate forests, by comparison, are likely to have fewer than 20 species in one 2.5 acre plot.

Even a backyard ten minute ecologist with access to a public library, a good bookstore, and a friendly commercial nursery could figure out the species diversity in a couple of neighborhood city blocks and compare that diversity to the 300-800 tree species found by the tropical forest biologists. My guess is that in the United States, even the most upscale neighborhoods are rather impoverished when viewed in terms of their botanical diversity. Similarly, or perhaps partly because of the impoverished plant community, animal diversity in most neighborhoods is very low. Suburban back yard animal diversity can be increased by keeping a rotting woodpile (I suggest sugar maple for best results, although that pile will certainly attract termites, at least in certain parts of the country), planting flowers that attract insects, and setting out a bird bath and bird feeders. But a year counting species from your kitchen window will still reveal only a tiny fraction of the backyard biological diversity you'd find if you lived outside of Manaus in equatorial Brazil.

Remember that any one species represents a unique set of genetic information and that this information, taken collectively, considering all species, is what spells "Life on Earth" (see chapter 2). About 70% of the Earth's genetic information resides in the tropics, or at least did prior to the largely successful post-war wholesale efforts to destroy it. Most biologists who work in the tropics immediately face a bewildering diversity of organisms, many of them nearly inaccessible to humans. In the latter category, for example, are the insects that live in tree canopy high above the forest floor.

Taxonomists, those whose work focuses on the discovery, description, and classification of species, tell us that there are at least 400,000 known species of beetles alone. Insect specialists who work in the tropics tell us, based on the rate at which they discover new species, that there may easily be a million species of beetles, and that they wouldn't be surprised to eventually discover several million, given enough time and talent applied to the task. Of course there is not enough time or talent to deal with the beetles, let alone the flatworms, roundworms, flies, snails, and various groups consisting of obscure yet often beautiful organisms.

Tropical complexity is not limited to the forests. In the warmer parts of the planet we also find coral reefs, which are almost in the same league as the tropical forests in terms of heterogeneity. Reef-building corals are members of the fascinating group known as Cnidaria, which includes sea anemones, jellyfish, Portuguese men-of-war, and all their relatives. Coral reefs are characteristic of warm shallow seas, mainly because the animals that secrete the calcium carbonate reef skeleton derive much of their energy from one-celled algae living in their tissues. These algae live inside the coral animal's cells and carry out photosynthesis (see chapter 7); the products of this process are then used for energy by the animal.

In this relationship, the algae get a home; the corals get energy; the mutually beneficial relationship is known as "symbiosis" (*sym* = together; *biosis* = living). For this reason, fossil corals are indicative of ancient shallow warm seas. Reefs support a great diversity of fish, worms, sponges, crustaceans, snails, and other invertebrates, but like many if not most habitats, the animal life of coral reefs includes vast numbers of microscopic species. And, of course, even microscopic creatures have their own unique life histories and relationships with their environments.

Tropical deforestation is a global ecological issue for a number of reasons, one of which is the rate and extent of this human activity. People clear forests for agriculture and housing, period. During the last century, for example, the great forests of the upper American Midwest were cleared completely by loggers who thought the timber supply was inexhaustible. Nowadays, of course, we have bulldozers and chainsaws, although in human hands fire—a factor that's been with us since the Earth was formed—also becomes forest-clearing technology. Population pressure and commerce increase the rate of deforestation.

It's not at all obvious what the ultimate effects of tropical deforestation will be; predictions range from global climate change to political instability and economic collapse. If it's the latter, then repercussions will certainly be global whether they involve climate or not. Strange as it might seem, the molecular biologists are worried about tropical deforestation, mainly because of the potential loss of genes which might in turn be the basis for new pharmaceuticals. For all these reasons, ten year ecologists deplore the destruction of tropical forests and work hard to slow the process. Whatever the long term effects, children being born this year are likely, within their lifetimes, to witness them.

11. Why is the Arctic so fragile?

> *In the stomach of a walrus butchered on the spring ice you will find the sediment of the ocean floor.*
>
> —Barry Lopez (*Arctic Dreams*, 1986)

By "fragile" I mean easily disrupted and slow to recover from a large disturbance. I gave this chapter its title mainly because of media attention paid to things like the *Exxon Valdez* oil spill in Prince William Sound, the Deepwater Horizon oil blowout in the Gulf of Mexico, and holes in the ozone layer. At the time of the *Exxon Valdez* incident, the newspapers were filled with dire predictions about the environmental consequences of not only the grounding and leak itself, but the entire Alaskan pipeline project. Pictures of oiled birds added to our image of the destruction, and news video of tedious beach cleaning supported the impression of serious damage to the ecosystem. Fishermen interviewed on television were angry and pessimistic about their futures.

At the other end of the globe, ozone depletion ("the hole in the ozone layer") was at first most obvious in the high southern latitudes, and again, one read various predictions about everything from the negative effect on global fisheries to increases in skin cancer. I have no intention of trying to settle any such ecological issues, especially in this book. But the polar regions have served historically as a major source of natural products, while remaining quite for-eign to most American readers, so any ten minute ecologist needs some explanation of polar biology.

Virtually all ecosystems, including those at the ends of the Earth, are fairly resilient and tend to recover from natural disruptions such as fires, floods, earthquakes, and landslides. "Natural" in this instance means "caused by something other than human activity." The speed of such recovery varies, however, as does our perception of it, and as you might suspect, recovery time also depends on the nature and extent of disturbance. In considering natural disasters, it's always helpful to remember that 50 years is a fairly long time to modern humans. Indeed, people starve to death and die of exposure in a matter of days or weeks, and national economies can easily fail over five-year periods. But to the planet, 50 years is only an eye-blink, if that, and a thousand years is not very long, either. So "easily disrupted" and "slow to recover" are relative terms, like many other ecological descriptors,.

Biological processes are generally somewhat temperature dependent and temperatures at the poles are low. In plain words, under cold conditions, such as in your home freezer, organic materials like leaves, bone, and skin, don't rot as quickly as they do in the tropics (cf. frozen mammoths periodically discovered in Siberia, or the Inca sacrificial victims' frozen mummies released from ice which is in turn thawed by volcanic activity). When an ecologist says "don't rot," however, he or she is really saying decomposers such as fungi and bacteria that dispose of dead bodies and organic waste don't grow very fast or luxuriously in the Arctic. Other biological processes, however, such as the growth of diatoms and crustaceans that feed on them, as well as the breeding of birds and whales that eat the crustaceans, seem to operate at fairly high speed—in Earth terms—if left to their own devices. So it's not really correct to say the Arctic is fragile because it's cold.

Before going much further, I need to point out again that the planet has two polar regions, the Arctic (north) and the Antarctic (south), a piece of rather common knowledge that, in an age of general geographical ignorance, no writer should take for granted. The Arctic circle is located 66 1/2° north latitude, and the Antarctic circle is located 66 1/2° south latitude, with the equator being a 0° latitude; thus the term "high latitudes" for the colder regions of Earth. Compared to the Antarctic, the Arctic is a veritable garden, primarily because the Antarctic is dominated by the continent of Antarctica, whereas most of the northern polar region consists of ice overlying ocean. Water under the North Pole retains far more heat than does the land under the south pole. Thus Arctic summers are relatively mild (!), but "Antarctic summer" is an oxymoron.

Strange as it might seem at first, in the summer daily temperatures fluctuate less at the North Pole than further south into Canada, mainly because the sun is above the horizon virtually all the time and there is little heat loss at "night." Similarly, in winter when the sun is constantly below the horizon, there is little daily temperature fluctuation because there is no daily source of heat. Antarctica is just plain cold; in the winter temperatures range from -40° to -100°F, although mid-summer temperatures can warm to freezing (32°F). The history of polar exploration is really a history of human beings venturing into brutal environments that stretch our vision of ecological diversity on Earth.

Life at high latitudes includes some rather characteristic plant forms, too, for example lichens, mosses, and stunted trees. Lichens are not really plants, however, even if they are included in some botany books, but instead double organisms, i.e. associations, between algae and fungus. Lichens also grow slowly under the best of circumstances, so slowly, in fact, compared to the more cooperative plants such as rhubarb and begonias, that humans easily lose patience with trying to watch lichens grow, and conclude they're dead. If you don't know what lichens look like, go outside to the nearest large tree; there's a fairly good chance some small ones are growing on old trunk bark. Take a magnifying glass along in order to appreciate the beauty of these lichens. Lichens in the Arctic and Antarctic, however, are ground dwellers and often much larger than ones on your local tree trunk.

True mosses, in contrast to many things people often call "moss," are mostly small plants that grow in clusters and as a rather continuous carpet, typically in moist areas. Mosses, and their relatives the liverworts and hornworts, lack the tissues known as xylem and phloem which transport fluids, thus don't have the kinds of internal structure as trees and shrubs. The tundra ecosystem—flat, wet, and cold—dominates the far north (alpine tundra occurs at high elevations even at temperate latitudes). The permafrost line is about 60° north latitude; above that line the deeper soil never thaws so water cannot drain down from the surface. There are nearly a thousand species of plants that live in the tundra, including fairly familiar ones like spruce and willow that occur not as upright stately trees, but as stunted, even creeping, forms nevertheless recognizable to those who know what to look for in order to identify a plant species.

In Antarctica, about the only places for plants to grow are rocky windswept ledges. This lack of space, coupled with the hostile environment and extensive ice cover, contribute to the rather low diversity of plant life in Antarctica, where about 400 kinds of lichens and fewer than a hundred kinds of mosses have been identified. Four hundred kinds of lichens seems like a lot of kinds to the average person, but on a global scale, four hundred kinds of anything is not very many (remember there are about 400,000 known kinds of beetles and almost a thousand kinds of bats). The northern polar regions are much more hospitable to plant life than the southern, and the Arctic flora includes a couple of thousand species of lichens, about five hundred kinds of mosses, and nearly a thousand kinds of flowering plants. But neither the Arctic nor the Antarctic are regions of high biodiversity compared to the tropics. Ecologists generally believe that stability is associated with diversity in natural systems. So relatively low biodiversity of polar regions leads one to predict that the stability of these ecosystems might be easily disrupted.

On the other hand, the Arctic oceans are, in many ways, highly productive environments. At the bottom of the food chain are algae, for example diatoms, brownish or golden one-celled algae that are exceedingly common throughout the world and grow on the underside of Arctic ice. In the Arctic summer, diatoms become part of the *phytoplankton*, a suspension of algae that flourish in the open ocean, especially at the surface. This mass of vegetation (an oceanic pasture!) in turn supports large populations of small grazing animals, for example shrimp-like crustaceans (krill) and other invertebrates, known collectively as the *zooplankton*. Among the zooplankton one finds a diversity of small animals, e.g. snails, jellyfish, comb jellies, arrow worms, many no longer than your thumb, that eat mostly one another.

Zooplankton is the main food for baleen whales, for example bowheads, which can easily eat a ton of this shrimp salad a day by straining it out of the water. Fish and squids also consume zooplankton, but then become prey to seabirds, seals and porpoises. The polar food webs are thus characterized by massive populations of a relatively few types of organisms. Polar food webs also have some seemingly unlikely connections, e.g. in the Antarctic between blue whales, the worlds largest living animals, weighing up to 150 tons, and their primary prey, 2-inch crustaceans (krill). A blue whale eats 3 tons of krill a day.

One doesn't usually think of 150-ton blue whales, or even smaller killer whales and penguins as being particularly "fragile," in the commonly used sense of the word. But, the generally low diversity of polar flora and fauna (plants and animals), compared to the tropics, these direct feeding connections between small and large organisms, and the large populations of consumers supported by standing crops of one-celled organisms and krill, all conspire to make polar ecosystems fairly vulnerable to disruption.

For example, global, and even cosmic, events that alter the amounts and qualities of light falling on the sun-starved poles can have profound effects on large standing crops of algae, thus also on shrimp, fish, penguins and other seabirds, seals, and whales. Such system-altering events can range from volcanic eruption to sunspot cycles. Phenomena such as holes in the ozone layer and greenhouse gasses, considered products of human activity, can also alter the quality of light falling on polar regions. This last observation is the basis for ten year ecologists' concern over effects that our daily activities have on Earth's ecosystems, especially polar ones.

As an aside, from my personal encounters with very many scientists over the past 50 years, it's my impression that the poles share one property with the deserts, namely the ability to simply capture the minds of people who go there to study nature. Of all the ecologists, geologists, and parasit-ologists I've known, those most wedded to their natural environments are those who've committed themselves to study of desert and polar ecosystems. I have no idea why this connection should be so, or even whether my claim would be born out by careful statistical analysis. But harsh environments have a way of reminding us that regardless of our tendency to place ourselves on a pedestal, we are, after all, just temporary visitors to one planet in a seemingly infinite universe. Scientists who study natural systems just seem to need that kind of reminder periodically, if for no other reason than to reinforce their values. This subject is dealt with in more detail in the chapter entitled "Why are ecologists such nerds?"

12. Why study islands?

Our knowledge is a little island in a great ocean of non-knowledge.

—Isaac Bashevis Singer
(*New York Times Magazine*, December 3, 1978)

Ecologists study islands for a number of reasons. First, many islands are beautiful and exotic places to work. Many are also miserable and isolated places to work, but working on them gets one away from telephones, committee meetings, babbling TV sets, cutsie e-mail and text messages, and annoying FAX machines. Second, some islands bear special relationships to the nearest mainland, and for that reason are exceedingly interesting places biologically. Third, islands are sometimes "living laboratories," in that their faunas and floras are unique and relatively impoverished, thus can tell us much about how organisms evolve in the presence or absence of forces that work on mainland relatives. Fourth, some of the best known ecological research was done on islands, spawning a whole sub-discipline of biology known as "island biogeography."

There are functional islands everywhere, i.e. habitats that differ qualitatively from their surrounding environments. A good example of the latter kind of island would be a cattail marsh in the middle of a prairie. And finally, some theories about space needs of plants and animals, developed during the study of island biogeography, are being applied to conservation problems in altered landscapes, e.g. to answer questions such as: How large must a wildlife refuge be in order to actually function as a wildlife refuge?

Perhaps the most fundamental reason for studying islands, however, is that much of the natural environment is distributed into patches of varying sizes—individual trees, lakes and ponds, fallen logs, the space under rocks, caves—and all of these kinds of places can, theoretically, be considered islands. Thus fish in a pond can't live in a nearby tree, so the pond is really an island inhabited by fish that are stuck in their "island." As is so often the case in science, a study conceived as a fairly narrow one, designed to answer a basic question, turns out to have rather broad application. The work that spawned the body of theory called "island biogeography" was done by Robert MacArthur and E. O. Wilson on island species diversity. Like many important theories, this one has kept at least one generation of ecologists busy investigating its predictions and applying them to cases far beyond those islands that MacArthur and Wilson studied.

There are some very important ideas about islands that have been revealed by research on them. One of the most important, and best known, is called the "area—diversity relationship." This idea can be expressed fairly simply as: *Large islands have more species on them than do small islands.* Ecologists studying islands, no matter whether the islands are glamorous exotic ones (Trinidad, Tobago, the Florida keys, etc.) or muddy mundane ones (farm ponds in the prairies), always consider the numbers of species, and the kinds of species, present as a basic part of their data. Large islands tend to be more complex (varied) ecologically than small ones because they have more space for different kinds of habitats to develop. A large island might have mountains, several different sizes of streams, forests, and coastal marshes, whereas a very small island generally does not have such a variety of habitats. This relationship between habitat complexity (variety) and diversity of life—namely that complex habitats support relatively diverse plant and animal communities—is actually a fundamental principle of ecology. And, this principle was developed most clearly by those studying islands.

Islands also differ in the rate at which they get colonized. The Galapagos Islands, Madagascar, and Hawaiian Islands are extraordinarily interesting islands and research on them has greatly influenced our views of how islands get populated. How do islands get populated? By a variety of means, although the mix of immigrants depends on a number of factors, most important among them being a suite of properties we call "vagility." The term "vagile" means free to move about, but to an ecologist, vagility refers to a species' inherent traits that provide motility in the geographic sense, i.e. over long distances. No matter how vagile a species may be, however, it still must cope with distance (between island and mainland), sometimes ocean currents, and prevailing winds.

Many birds, of course, are quite vagile because they can fly anywhere. Barn owls and barn swallows are so vagile, for example, that they occur naturally almost everywhere on Earth. Some spiders also are highly vagile because as microscopic newly hatched spiderlings, they spin out long strands of web that get caught by the wind and carried long distances. Aquatic insects such as whirligig beetles and backswimmers are also fairly vagile; in one memorable instance a backswimmer appeared one morning in a half-empty drink glass left on our deck after a party the night before. Cleaning up the morning after and finding a backswimmer in a glass on your back deck makes you think seriously about the various ways islands can be colonized. Freshwater fish tend not to be so vagile as backswimmers, but humans have increased the vagility of many fish by carrying them around in minnow buckets, releasing them, or introducing them into new islands of water as part of some wildlife management scheme.

Because islands are often more difficult to colonize than half empty drink glasses, the mix of species on islands is typically different from that of the nearest mainland, thus the evolutionary forces acting on island species are different from those acting on related species on the mainland. Gigantism and flightlessness are two phenomena that evidently a-rise in island fauna when the pressures of predation are removed. Giant tortoises occur in a number of places in the world—the Galapagos, Seychelles, and Maldive Islands, for example. These islands are distantly separated, and the tortoises that originally colonized them were from different genetic stocks, so that the gigantism evolved independently in the different locations.

Flightlessness among birds also occurs on protected is-lands. One finds flightless cormorants on the Galapagos Islands, for example, but the best known product of island isolation was the dodo, a giant pigeon. Large, slow, flightless, great-tasting animals tend to disappear when predators, especially human ones, arrive on islands. In the case of giant tortoises and dodos, human sailors did the job. The dodo is extinct—killed and eaten mainly by sailors—and giant tortoises are uncommon. Sailors, including those aboard the *Beagle* on which Darwin sailed around the world, took Galapagos tortoises along as food because they lived just fine on ship and didn't require refrigeration.

Lake Titicaca, high up in the Andes Mountains, also functions like an island, evolutionarily speaking, and a flightless cormorant (large fish-eating bird related to pelicans) has evolved there, too. Anyone who's watched cormorants for very long knows that they don't fly much anyway, and are especially disinclined to fly when there are lots of fish and other cormorants around, the weather's reasonably comfortable, and nobody's shooting at them. So the evolution of flightlessness in cormorants should not be too much of a surprise, even to a ten minute ecologist.

Lake Baikal, covering about 12,000 square miles in southern Siberia, is the world's deepest lake and also one of the most isolated (= insular!). Lake Baikal is another of the world's famous islands, and it is filled with unique fishes that have evolved there, in a manner analogous to the evolution of unique life forms on the island continent of

Australia. Lakes Titicaca and Baikal are aquatic islands in seas of land. The prairie mud-holes mentioned earlier are just tiny versions of those famous lakes; but mudholes are rarely big enough, and don't last long enough, for unique species to evolve in them, so tend to maintain their species diversity by drink-glass type immigration.

The theories of island biogeography are important in a practical sense because they focus attention on the areas and qualities of preserves. Thus these theories are an essential part of our resource management strategies, and especially conservation, schemes. How big must a wildlife refuge be? is a question commonly asked by both politicians and conservation biologists (although perhaps for different reasons). The answer is: Large enough to actually provide a refuge (if that's the intent).

How large is "large enough?" Ahh! That is a very good question, and its answer depends somewhat on the species involved. Whatever the species, however, a refuge must be large enough to dilute what's called "edge effects." For example, a woodland designed as a refuge for birds that nest in the woods must be large enough to damp the effects of nest predators and social parasites that naturally forage mostly along edges of clearings. A good example of cruising social parasites are cowbirds that lay their eggs in other birds' nests. Small woodland birds, such as warblers, that nest near the edges of clearings end up rearing mainly baby cowbirds. So unless the forest bird sanctuary is big enough to provide nesting sites far from the edge, the patch of woods is not much of a preserve.

Thus "habitat fragmentation" is a major issue with conservationists who see breakup of habitats to be as serious a problem as the loss of habitats. In fact, breaking up a once continuous habitat can be the equivalent of destroying it by, in essence, making islands out of a former sea (metaphorically speaking). Conversely, species that require islands, such as wetlands and potholes in a vast prairie, need enough of these kinds of islands to counteract the natural forces of drought, fire, and storms. Thus when the conservationists speak of trying to save a species, they're actually talking about trying to save enough of its habitat in whatever form—island or sea = patchy or continuous—the species needs in order to survive.

13. How is real estate really divided up?

Our soil belongs also to unborn generations.
—Sam Rayburn (from V.
S. Young's *The Speak-
er's Agent*, 1956)

Humans tend to divide up real estate according to a bewildering array of reasons and methods. The reasons range from hatred and desire to tradition and necessity. As a minimum, the methods humans use to divide up real estate include contracts, inheritance, and warfare. In general, these human reasons and methods don't apply to natural systems, although it could be argued that necessity and warfare, of a sort, play major roles in the establishment of animal territories and plant ranges. Instead, in nature, real estate tends to be divided up according to available moisture, temperature, wind, elevation, latitude, geological history, and soil type. In other words, the abiotic (non-living) forces that operate on the planet tend to be the primary factors that work to divide real estate into recognizable pieces.

These abiotic forces are not necessarily mutually exclusive. For example, elevation and latitude both influence temperature (the air gets colder the higher one goes up a mountain and the further north or south one goes from the equator). Soil type of a region also depends at least somewhat on the geological history of that region. So in essence, when it comes to real estate, Mother Nature doesn't give a damn what humans think or want. All these forces of nature conspire to divide real estate into desert, mountains, ocean, dry short grass prairie, coniferous (evergreen) forest, tropical rain forest, deciduous (trees shed leaves) woodland, and numerous other types of places. Humans then come along and try to divide up the already-divided real estate into nations, cities, farms, subdivisions, blocks, lots, and shopping malls.

But humans are not completely independent of natural forces. The agricultural economy of a particular region is dictated to a very large extent by the way Mother Nature divides up her real estate. The agricultural economy of the large coniferous forests is always most likely to be timber and furs. Similarly, the agricultural economy of those areas Mother Nature has set aside as grasslands is most likely to be grain and domestic stock production. This principle—natural phenomena dictate agricultural economy—is not very difficult to understand. Prior to the discovery of mineral wealth in various places (oil, coal, gold), the inhabitants of a region typically lived according to the prevailing agricultural economy. Often social structures and literature—religion, family rituals, entertainment, education, myths, songs, and stories—were, and to a great extent still are, strongly influenced by Mother Nature's real estate transactions.

Mineral wealth—an accident of geological history—completely alters the overall economic picture, as, for example, in the Middle East, where oil has temporarily changed a predominately desert agricultural economy into an economy based on the global price and supply of fossil fuel. In the process, the political importance of various Middle Eastern nations has been altered significantly because of the modern relationship between petroleum—no matter where it comes from—and social-political events, including war. Thus nations (human structures superimposed on natural real estate) that rely heavily on petroleum are likely to fight to keep their supplies. But without modern technology and mineral wealth accidents, economies are still tied mostly to Mother Nature's real estate transactions.

The word "biome" is commonly used to describe the various kinds of natural divisions in real estate. One hears of the "desert biome," the "alpine tundra biome," the "tropical rainforest biome," etc. The term "biome" is a useful one because it combines an enormous amount of information into small phrases or pair of words, then allows even a ten minute ecologist to communicate that wealth of information. Why is this true? Because specific biomes, e.g. prairie, savannah, or deciduous forest, are established on the basis of their dominant vegetation, which in turn is a reflection of the available moisture and temperature. The natural vegetation of a particular biome, the vegetation that dictates agricultural economies, is called the *climax vegetation*. Climax vegetation is the tallest, most water-demanding, vegetation that can grow in a particular area, given the long term supply of water and temperature.

Among the most classic of all ecological research is that done on "succession," or the sequence of plants and animals that occupy a disturbed area. For example, a plowed field is a disturbance, and the images we have of heroic pioneers and settlers, clearing the land and putting in crops, are actually mental pictures of people stripping off the climax vegetation that characterized a relatively stable biome. What happens when a cultivated field is abandoned? Virtually everyone knows the answer to that question: the field is repopulated by weeds, which then give way to more long lived plants, and eventually only experts can tell that there was ever a plowed field on that site.

What enables the experts to determine this bit of history? The experts we're talking about are actually plant specialists who can "read" the vegetation and recover the land use history from the mix of species and the ages of individual plants. A repopulation sequence—beginning with "weeds" and ending with longer lived plants—is called succession (because generally one type of vegetation succeeds another). "Weeds" are weeds primarily because they intrude into our lives whenever we decide to start a succession sequence, for example by digging up an intended garden. Weeds can also be thought of as hardy and opportunistic pioneers, a more flattering name than, well, *weed*.

Succession proceeds at different rates and to different ends, depending on where it occurs. Those rates and ends are a product of available moisture and temperature. There are some fairly famous pictures in the ecological literature. One is of a triangle graph that looks something like a mountain with moisture content increasing from left to right across the bottom, and temperature decreasing from hot to cold as one ascends from the bottom to the top. Obviously each point on this triangle is a unique combination of temperature and moisture. In this iconographic mountain-graph picture, various regions are labeled with their climax vegetation type according to these temperature and moisture combinations. I always thought that was a pretty instructive picture, explaining, in relatively simple terms, the origin of desert, deciduous forests, and tundra.

The other famous picture shows two views of the changing vegetation of North America, especially the United States and Canada, as one travels from east to west at a single latitude, and south to north at a single longitude. The east-west drive is at about the latitude of Interstate 40 (from the Carolinas to Bakersfield, California, roughly along the 36th parallel). The changing vegetation one sees on this trip is more a reflection of decreasing moisture than of different temperatures (because I-40 is about the same distance from the equator along most of its transcontinental length). At Raleigh, North Carolina, temperatures (Fahrenheit) range from the 50s in winter to the high 80s in summer, and these temperatures are combined with 45-50 inches of precipitation a year, including about 10 days of rain or snow a month. Climax vegetation is a mixed deciduous and pine forest (textbooks call it "temperate deciduous forest). At Bakersfield, California, winter high temperatures are also generally in the 50s, and summer highs are in the 90s, but that city experiences only about 3 days of rain a month, and a total of 8-16 inches of precipitation a year. Not surprisingly, textbooks call Bakersfield a desert.

There are no interstate highways laid out along the 100th meridian, so a ten minute ecologist wishing to experience the other half of this famous picture would have to wend his or her way northward along blue highways from western Oklahoma to Manitoba (where roads pretty much end a hundred miles north of Winnipeg). In places like Kingfisher, Oklahoma, temperatures range from the high 40s in winter to mid-90s in summer, with about 6 days a month and 25 inches of precipitation a year. At La Pas, the "jumping off place" for the great northern Manitoba wilderness, January highs are near 0° F, July highs are in the 70s, with almost 9 days a month and 15-20 inches a year of precipitation. Climax vegetation at Kingfisher, Oklahoma, is perennial grass ("mixed and short grass prairie"); at La Pas, one steps out of town into a boreal coniferous forest that pretty much circles the globe at the higher northern latitudes.

Current thinking about climax vegetation is that it may be more localized than once thought, so that within a region, e.g. the Great Plains, although the dominant vegetation may be perennial grasses, there will be places where trees take over naturally. There are also places where trees take over unnaturally, for example landscaped yards where exotics are watered regularly. Conversely, there are pockets of seemingly out of place vegetation types throughout much of the world, so that when one looks at vegetation maps, one sees a complex patchwork at small scales. But that patchwork is superimposed on the larger patterns described above.

Real estate also is divided up into continents, subcontinents, peninsulas, and islands, and if we consider bodies of water "real estate," then add streams, rivers, ponds, lakes, roadside ditches, rain puddles, fjords, bays, and oceans. In other words, natural physical spaces created by the distribution of land and water are as highly diverse in size and kind as are legal spaces created by contracts and deeds. Humans tend to dictate activities conducted in their spaces, although to some extent the reverse is true, but then humans also create legal spaces for the express purpose of dictating the activities; examples of this latter principle are artists' studios and church sanctuaries. The structure and location of natural physical spaces, however, have a profound effect on the spaces' occupants and their activities.

Awareness is the cheapest of all traits to acquire, being much less expensive than a college education, for example. Ten year ecologists are constantly aware of their surroundings and tend to interpret their observations in terms of their knowledge of nature. Ten minute ecologists can very easily acquire a significant measure of awareness simply by browsing through an atlas prior to travel, reading about natural vegetation in places where we and our relatives live, and taking advantage of botanical gardens when they're available. But be forewarned. Awareness of the natural world has a tendency to turn normal people into ecotourists (which some of us believe is not all that much of a disaster).

14. How many is too many?

. . . it is futile to expect a hungry and squalid population to be anything but violent and gross.

—Thomas Henry Huxley
(*Joseph Priestley*, 1874)

In most people's minds, the name Malthus is connected with the concept of overpopulation, and the adjective "Malthusian" refers to the dire consequences of reproducing to the point that we run out of resources. Charles Darwin, recognizing that in nature more individuals are born than typically survive, used this idea as one of the central themes of his evolutionary theories. Ecologists have known for a long time that populations of wild plants and animals fluctuate, often dramatically, and have studied many specific instances of this phenomenon in search of causes. The list of potential causes is fairly long, and includes everything from predators to sun spots. In addition, the ecological literature is filled with experimental studies, of many kinds of plants and animals, in which the causes of population fluctuations have been either demonstrated or inferred. In these cases some needed resource is usually shown to be a limiting factor on numbers.

Ecologists also have known for a very long time that the populations of many plants and animals are relatively stable but limited. In other words, these populations do not necessarily undergo large fluctuations, nor do they increase regularly. Classic examples of such populations include the vast evergreen conifer softwood forests of North America and Eurasia, which is the dominant plant community of the cool, relatively moist, but stable, climate occurring throughout much of the world between latitudes 45° and 65° north (those latitudes in the southern hemisphere are occupied mainly by ocean). Indeed such forests are so characteristic of these latitudes that we have a name for them: boreal forest, or taiga. The dominant plants are spruce and fir. We don't see boreal forests spreading down over the American Great Plains or the Russian steppes every year, or covering the polar ice cap one year then retreating the next. Instead, we consider them characteristic of the north, and stable enough to harvest for wood without making much of an impact on them. The first of these considerations, namely their natural geographic distribution, is certainly correct.

In these vast conifer forests literally billions of seeds (in cones) are produced annually, but only an infinitesimal fraction of those seeds ever produces a mature, seed-bearing, tree. Most of these seeds end up in places where they won't germinate, for example in bird and rodent stomachs. So most seeds die, but some end up in places where they survive without germinating. Then under correct conditions of moisture and temperature, they germinate. These seeds are like money in a bank, a seed bank, just waiting for the right time to sprout, a sort of natural savings account.

Under natural conditions that prevailed throughout most of North American history, at least the last fifty million years of it, fires, floods, and death from disease or old age, all provided an opportunity (space, time, unoccupied soil) for some of the seed bank seeds to germinate and grow to old trees. Available moisture and temperature, produced by global climate patterns, determined the geographical extent of the spruce and fir populations. Local soil and water conditions produced small scale fluctuations in these populations and affected survival of individual plants.

In a grand sense, therefore, prior to the arrival of humans on the North American continent, the evergreen conifer forest populations were both limited and regulated by naturally occurring events. That is, these trees lived according to Malthusian rules. The key to understanding this fact is our ability to appreciate both the time and space scales involved in the limitations and regulations. Of course it also helps us understand population regulation if we can depersonalize the subject, which is why I started out discussing the vast forests most of us never see, and think of (when we think about conifer forests) mainly as a source of timber.

That is, most of us don't consider the Malthusian aspects of spruce tree lives to be a moral or religious issue. The only time we ask whether there are too many spruce or pine trees is when we are talking with a landscape architect. But obviously, throughout thousands of years of history, long before there were landscape architects, natural forces were determining that there were too many spruce and pine trees and confined most of the survivors to northern latitudes.

"Too many" is clearly a human idea, and it refers to the numbers that can be supported by a particular set of resources. Nature really doesn't care whether organisms live or die; only humans care. But we can explain "too many" in a rather neutral and practical way by considering what ecologists call "carrying capacity." Carrying capacity is the number of organisms, of a particular species, that an environment can support. Two environments might have different carrying capacities for the same species; one environment might have different carrying capacities for different species.

As far as nature is concerned, "too many" actually means "a higher number than the carrying capacity." In the northern forest example given above, when more seeds are produced than can ever germinate and grow into an adult, then the dead seeds are those individuals produced beyond the carrying capacity of the biogeographical region we know as boreal forest. A lightning-strike fire, however, can provide a local opportunity (clear a place) for more new trees (germinated and grown seeds) than could otherwise grow in that spot. But the burned area still has only a certain carrying capacity, and once that capacity has been achieved, the population will be relatively stable and disturbed only by death.

This general principle of the relationship between population size and environmental carrying capacity is one of the most truly fundamental and broadly applicable ecological concepts. Indeed, it applies not only to boreal forests, but to individual bacteria in a test tube. Speaking of test tubes, it's time to tell you my favorite story, namely the one about the talking bacteria. These bacteria divide once every minute and they live in a test tube that has only enough food and space in it to support this kind of reproduction for an hour. A little bit of simple math helps understand what these talking bacteria are saying.

A minute after the scientist puts one bacterium into the tube, let's say at 11:00PM, there are two bacteria, and after another minute (two minutes into their hour), there are four. They have enough culture medium to keep reproducing at this rate for an hour, so at midnight, of course, these remarkable bacteria will run out of resources and die. Calculating backwards, we discover that one minute before midnight, the tube is half filled with bacteria, and two minutes before midnight, it's a fourth filled (= 3/4 empty). So from 11:00PM until 11:57PM, these bacteria have flourished, multiplied, and diversified into a very large community that includes bacterial politicians, businessmen, ecologists, and of course college students (remember, it's only a story). But at 11:57PM this community can exist for only three more minutes.

And what do these talking bacteria say at 11:57PM? The ecologists, of course, say what ecologists have been saying for quite a while, namely that *we're about to run out of resources*. The politicians, likewise, say what politicians have been saying for quite a while, namely that *the ecologists are idiots; we have three times as many resources as we've used throughout all our history, so don't worry (and vote for me because I'm so smart)*. In a similar manner, the bacterial businessmen say what businessmen have been saying forever and forever, namely that *our political leaders are right; we have three times has many resources as we've used throughout all recorded history, so we should sell some to another test tube (and give me a tax break to create the new jobs produced by these sales).* And the bacterial college students are asking what young people ask all the time, namely: *Who should I believe?* Personally I think we should listen to the ecologists; but I'm somewhat biased.

Obviously the tale of the talking bacteria is only a teacher's device for putting the concept of "too many" into various perspectives. I'm not entirely sure where I got this story, but it may have come from a physicist at the University of Colorado, Dr. Al Bartlett, who made a career out of analyzing our resource use. If this is where the story came from, I acknowledge and thank Dr. Bartlett. I may also have mentioned it in some form in another book I wrote, and if so, acknowledge both Dr. Bartlett and the other book (in which I also acknowledged Dr. Bartlett's contribution to my numerical awareness).

The point of this story is fairly obvious, too: no matter what you want to believe about the natural world, we are still very much a part of that world, and there are certain fundamental ecological principles that operate on all organisms no matter what they believe. One of these principles is that environments possess carrying capacities and will not support populations larger than those capacities, no matter what politicians and businessmen claim. The other principle is that exponential growth brings populations to their environmental carrying capacity at an increasing speed. As the talking bacteria are about to discover, in periods of exponential growth, it's fairly easy to delude oneself about just how rapidly a population is approaching its limits.

To what extent do these principles apply to humans? That is a good question with many answers. Mathematicians, as well as most college students, can easily calculate the year at which the mass of humanity comes to exceed the mass of Earth, assuming that human reproductive rates remain what they are today. It doesn't take a rocket scientist, or even a ten minute ecologist, to figure out that some time prior to that date, humanity will begin to live a rather Malthusian existence. On the other hand, something may happen to stop or slow human population growth well before it reaches the Earth's carrying capacity. Malthus thought that "something" was a combination of war and disease. Many people today believe that the "something" is technology, which won't necessarily stop our population from growing, but will greatly increase, perhaps even to near infinity, Earth's carrying capacity.

Most scientists believe rather strongly that such faith in technology is perhaps not justified. Most scientists have seen too many experiments fail, had too much experience with technology that simply didn't work like it was supposed to, etc., to ever place complete faith in that ethereal and unpredictable phenomenon known as "technology." Most scientists, and especially most ecologists, also understand that simply being alive is not necessarily the same as being human. Thus humans appear to need intellectual stimulation—art, music, literature, science, a few problems to solve, meaningful work, etc.—in addition to bread and milk. And, most ecologists understand clearly that although we may run out of bread, milk, art, and literature, we have a seemingly inexhaustible supply of problems to solve. These same ecologists also know that the problem of maintaining our humanity as we near Earth's carrying capacity is the most difficult of all this inexhaustible supply of challenges.

15. How long is short? (or, How short is long?)

> . . . like the ever-flowing stream of time, the
> beginning and the end.
> —Rachel Carson (*The Sea
> Around Us*, 1951)

For this chapter, "short" and "long" will refer to time rather than size. In chapter 1, I characterized humans as being short-sighted, and hopefully portrayed that characteristic as a rather natural one, of biological origin, rather than a condemnation of attitude and behavior. I don't believe humans necessarily must act against their own best interests, however, just because they happen to have inherited a trait from their ancestors. On the other hand, humans don't always understand that "short" and "long" are relative terms, not absolute ones. For this reason, what might seem like a long time to us may be, to the planet, a virtual eye-blink, and what might seem like a short time to us, can be an eternity to a microorganism.

Ecologists typically use the term "scale" to describe their point of reference. Events that occur over the course of a minute are small scale ones in comparison to those that occur over the course of a day. Similarly, events that occur over a ten million-year period are large scale ones in comparison to those that occur over our lifetimes. Ecologists, and indeed most scientists, shift their thoughts easily along these sliding relative time scales. In fact, if I had to pick one trait that separates ecologists, and organismic biologists in general, from the rest of humanity, it would be this trait of being able to direct their thinking along several relative time scales.

Thus one secret to acquiring ecological knowledge and insights is learning how to shift one's thoughts easily from today to yesterday to the middle of the next century then back to the Pleistocene. The minds of molecular biologists, those who study, and often manipulate large molecules such as DNA, typically don't move so easily among time scales from minutes to hundreds of millions of years, unless they're using molecular technology to study evolution. But ecologists, searching for explanations of the numerous phenomena observed in nature, must think in terms of everything from prey-predator relationships to evolutionary histories. Thus it should be no surprise that ecologists have adopted many of the powerful molecular techniques to address their questions about the lives of organisms.

Acts of predation, as anyone who's watched many nature shows on television knows, occur over very short time scales—seconds, usually, minutes or hours at most if one includes stalking or waiting for prey. Evolutionary histories, however, are played out over long periods—days to months in the case of viruses, months to years in the case of bacteria, decades in the case of some plants, and millennia for most of the rest of us. An ecologist studying the evolution of predatory behavior must, of necessity, compress both time scales into a single package of understanding. I don't believe ecologists are born with such abilities; like other practicing professionals, ecologists have to learn the tricks of their trade, and this time scale vision is one of those tricks.

Long and short times are also important considerations when dealing with questions of stability. Governments, cultures, religions, school athletic programs, businesses, and indeed most human organizations are concerned with stability. But stability is simply not the rule in living systems, including governments and football teams built by humans. Nor is stability the general rule, at least over the long term, in nature. Aha! The words "long term" again raise the issue of scale. At the longest scales, we can say that no populations are stable. The vast and overwhelming majority of species that have ever lived on Earth are now extinct. But on shorter time scales, we often see that populations are often relatively stable, being "regulated" by various natural processes. Such regulation always involves dynamic, give-and-take interactions between organisms and their environments. Sometimes such interactions are highly complex, with multiple factors at work. So population regulation, like all things that are dynamic and complex, are interesting to ecologists whose entire business is characterized by dynamism and complexity.

Dynamism and complexity are also interesting to those whose business is real business. For this reason, as well as some other, more compelling, ones, ecology and economics tend to share certain ideas, although these ideas may not be expressed in identical terms. "Supply," "capacity," "growth," and "optimal," are all terms that apply both to ecology and business. Business people know that short and long term profitability are not always strongly connected. Suppliers ("supply"), markets (as in "saturated market," describing "capacity"), quarterly profits (calculate to determine "growth"), and the relationships between these factors (are the relationships "optimal?"), are ongoing concerns for business.

Similarly, populations of roadside daisies are regulated by their supply of moisture, the capacity of a ditch to satisfy their requirements, the growing season, and the presence of other plants. And everyone knows that whatever happens to these daisies in the next few days may or may not determine what happens to them in the next few years. Ecologists try to understand the time scales, dynamism, and complexity of events that determine whether the daisies are in the ditch tomorrow, and whether daisies will still be growing beside that road when his or her children are old enough to see them. Business executives, too, try to understand the events that determine how long the business will operate, and whether their own children will have an opportunity for meaningful employment.

Ten minute ecologists need to pay attention to issues of scale, primarily in order to understand the various resource utilization arguments swirling around in the political arena. The economic term "growth" is probably the most important concept to apply to public discussion of ecological issues. To "grow" is to get larger, thus "growth" refers to a *process*, not an active verb as we often hear nowadays in political discourse. Anyone who's reared a child, or bought a baby puppy (or kitten) and watched it grow, knows that a 12 pound dog "grows" into a 15 pound dog, and then (that same) 15 pound dog grows into a 20 pound dog, etc. The same rules apply to kids, kittens, earthworms and elephants. In other words, growth is added on to the previous size.

Percentage growth, by which economic health is often assessed, is a very misleading indicator of anything, especially economic health. A 12 lb dog that grows 5% in a month adds 0.6 lb; a 15 lb dog that grows 5% in a month, however, adds 0.75 lb, and a healthy, growing 20 lb pup puts on a whole pound, all during the first month. Now, if we're talking about the same puppy, growing at the rate of 5% *per* month, then that animal is putting on *increasing amounts of weight (= numbers of pounds) every month*. If your 20 lb puppy keeps growing at the 5% *rate*, then at the end of another year you'll have a 36 lb dog. If it has just kept putting on a pound a month, however, you'd have only a 32 lb dog. Percentage growth equals compound interest, and over the course of the year it gave you nearly 40% more puppy than you'd have had if it had simply added a pound every month.

In most cases, over long time periods, percentage growth of various processes eventually brings the growing organism up against its environmental limits. Why? Because some essential resources—water, energy, food of a particular kind—are themselves typically limited, and these resources rarely if ever grow, especially not in percentage terms. If your puppy keeps growing at 5% per month, then after its second year it will weigh 65+ pounds and be a rather hefty playmate. But if it continues growing at that same rate for another ten years (typical life of puppies), then you've got a six ton dinosaur dog curled up beside the fireplace or roaming the neighborhood. Obviously something in most puppies' genes stops them from growing indefinitely.

Businessmen and politicians, however, tend to view continuous percentage growth as a sign of a healthy economy. Ten minute ecologists know different because they are able to shift their thoughts and analyses between both short and long term time periods. In their minds, the cute baby puppy that finally got house trained eventually grows into the six ton monster that has to be fed, housed, leashed, and made to heel and fetch a ball if it's to be kept "healthy."

Ten minute ecologists also transfer ideas and concepts easily from one situation to another. Percentage growth in utilization of a limited resource is one of the more easily transferred concepts. Over the long run, such utilization will inevitably bring the user up against the harsh reality of limits, whether the resource be fresh water or petroleum. But over the short run, we usually can get by with percentage increases in consumption of limited resources.

As a political aside, I've always thought it strange that people seem to have such fear of things perceived as wastes (nuclear and otherwise) and such a cavalier attitude toward resources. Thus "not in my back yard" is a rallying cry for opposition to everything from halfway houses to landfills to nuclear power plants, while "not in my lifetime" [will they run out] is the usual excuse for continued percentage-growth exploitation of fixed resources. Thus a radioactive isotope with a half life of several thousand years inspires fear, even though the level of that isotope may be comparable to radioactivity in a brick house, while an essential resource with a whole life of a couple of hundred years inspires denial.

This may be a good time to consult the previous chapter again, in which the talking bacteria discussed their various perceptions of growth against the limited resources in their test tube. Most ecologists view Earth as simply a big test tube. And, in a sense, these scientists are correct. After all, there's no new Earth filled with fresh water, abundant petroleum, and vast continents of arable land for the taking. There's only the old Earth, that's been here for five billion years and watched a long parade of life forms come and go, some, such as sea stars (starfish) and their relatives, lasting for several hundred million years and surviving mass extinctions, and others, such as humans, existing for only a fraction of that time (so far) and remaining untested by giant meteorites and continent-rending volcanic eruptions.

In summary, one trick to becoming a ten minute ecologist is to learn how to interpret information in terms of various time scales. This task is not a very difficult one. All you have to do is play "what if?" with all the ecological, and indeed economic, information that comes your way via television, radio, newspapers, magazines, and the Internet.

16. What good is a swamp?

*But Mr. Jeremy liked getting his
feet wet; nobody ever scolded
him and he never got a cold.*

—Beatrix Potter (*The
Tale of Mr. Jeremy
Fisher*, 1906)

A swamp is good for many reasons, although people
who are not ecologists don't often think in terms that make
these reasons valid ones (for them). Also, "swamp" can
mean many things to many people. The easiest way to define
"swamp" is to equate it with "wetland." After all, a swamp is
a form of wetland, but not all wetlands are what we might
typically call swamps. Some wetlands are marshes. My dic-
tionary defines "swamp" as "a wet spongy land saturated and
sometimes covered with water." "Wetland" is described as
"lands (such as tidal flats or swamps) containing much
moisture." So I believe our equation of the two has some basis in the traditional uses of our language.

The term "wetland," however, also has some fairly
serious political implications, in part because of the Federal
government's attempts to help protect them, but most im-
portantly because of government's attempts to define exactly
what they are, and are not. I'm using the term "swamp" in
this book because it's more fun than "wetland." "Swamp"
makes me think of wading into some mucky place, having
dragonflies circling around your head, finding wonderful
tiny organisms, and getting real dirty. "Wetland" makes me
think of politicians in 3-piece suits, Federal regulations, environmental extremists, and farmers with guns.

When the average citizen asks the question: "What
good is a swamp?" then that person is really asking "of what
economic value is this wetland?" This question is a very difficult one to answer, and our satisfaction with the answer(s)

depends on our perception of economic value. For most people, economic value is short term and immediate: a job that pays high wages, a piece of property that yields rich crops or valuable minerals, a gold ring, a Jackson Pollock painting, any specific item that can be bought or sold or provides money on a schedule that allows us to buy things like groceries, gasoline, and utility services. But ecologists tend to think in more abstract, general, and longer terms than average citizens. So when you ask an ecologist "what good is a swamp?" you're likely to get an answer like "insects are the glue that holds life on Earth together." Many people who get such answers respond with another question: H-u-u-u-uh?

Basically, swamps are "good" for several reasons, the most important ones being philosophical, educational, and spiritual. Humans tend to have a certain intellectual relationship with their surroundings, and those who live, or have access to, wilderness often acquire a certain reverence for undeveloped areas such as mountains, deserts, and, yes, believe it or not, swamps. Such swamp neighbors soon learn that swamps are "good" because:

(1) Swamps are places where untold numbers— literally billions and billions—of microscopic and nearly microscopic organisms live, breed, and die. These tiny organisms are near the bottom of the food pyramid (see chapter 7), and they are eaten by equally untold numbers of larger organisms, although "larger" in this case may be somewhat of a misnomer, because down at the bottom of the food chain, even predators are sometimes quite small. So, nature's terrestrial food pyramid, that is, the one operating on land, rests in part on the wetlands. Near the top of this food pyramid are birds, including migratory ones that utilize wetlands on more than one continent, eat flying, as well as non-flying, insects (read: mosquitoes and lawn grubs), and provide us with enormous pleasure (see #2 below).

Many aquatic food pyramids also rest on wetlands adjacent to rivers, lakes, and yes, even oceans. Wetlands bordering a lake provide shallow waters for young fish to hide in, and at the same time provide a massive crop of the small insects and crustaceans that newly-hatched fish eat. After all, a 10 pound bass starts out as an almost microscopic hatchling, which in turn must find lots of even more microscopic food, while at the same time avoiding its larger hungry cousins, if it's ever to grow into that lunker being held up so proudly in the smart phone photo by that grinning fisherman who's just spent $35 for a fishing permit, $250 for clothing, $175 for a depth finder, $80 for a local motel room, $100 for food in the past couple of days, $18 for a case of beer, $200 in gasoline for his sport-utility vehicle for which he paid $36,500 (plus taxes) so he could be a sportsman, and $5,700 for a boat, and is about to spend untold thousands of dollars, upon which lawyers will pay taxes, when he gets home to find out his wife's divorced him for spending so much time and money fishing with the boys. Maybe the SUV was the last straw.

(2) Swamps are patches of habitats where many migrating birds spend significant amounts of time. Significant may not mean weeks or months, so much as feeding time. A day in a swamp is worth a week in a less productive habitat for many birds. A good example of this dependence is the lesser sandhill crane that migrates through the Great Plains in the spring, congregating for a month or so in a hundred mile stretch of the Platte River in Nebraska, These cranes eat vast numbers of earthworms and snails produced in wet meadows and small "swamps" adjacent to the river. During March and early April, the skies are filled with dramatic flocks of these large birds uttering their primeval calls. The local tourist boom, produced by people from all over the world coming to places like Grand Island, Nebraska, results in a major economic benefit to the region, as well as traffic headaches for local law enforcement officials.

Scientists believe that weight gained by these birds in March, as well as the calcium derived from snail shells, are both essential to nesting success in the Canada. I cite the sandhill crane case because it's such a clear and obvious example of big-time economic benefits being derived from wildlife protection, conservation, and wetlands. The main reason it's so clear and obvious is because of the heavy duty advertising and crane souvenir business that goes on all year in that part of the country. But there are dozens, perhaps hundreds, of cases such as this one in which wetland based ecotourism is a truly major industry.

(3) Swamps are places where scientists, teachers, and students learn of the mechanisms by which nature replenishes itself and detoxifies our wastes. For this reason, to whatever extent basic knowledge about the way the planet operates is of value, that's the extent swamps are also of value. (Swamps are not alone in this regard, but they stand pretty high on the list of important types of habitat laboratories.)

Reason number 3 may be the most important by far, and it applies not only to swamps, but also for many other kinds of natural areas. A ten minute ecologist needs to remember that certain types of classrooms cannot be built. They must be found instead, and without those classrooms we cannot produce scientists of a particular variety. Swamps, mountain ranges, volcanoes, earthquake zones, coral reefs, prairies, rain forests, are all good examples of such classrooms. Without them we cannot produce the scientists that give us the information necessary to make rational choices about natural resource utilization. These scientists are the ten year ecologists. When it comes to long term ecological processes that affect our lives, ignorance is the most costly of luxuries, and knowledge, including that provided for researchers by readily available wetlands, is the most economical of necessities.

Generally, however, our accountants, politicians, and tax lawyers are not trained to think in such terms. Accountants, politicians, tax lawyers, and a few other non-scientific types also think that real swamps can be replaced by computer swamps. They can't; nothing demonstrates the stupidity of this belief more than a bucket of swamp gunk and a kid who's just been asked "What's all that stuff?" That's one of the reasons why I wrote this book, for this poor kid, his or her parents, and his or her teachers. This budding scientist, who eventually will contribute significantly to his or her nation's economic and political stability through research, teaching, and public service, is about to discover that the virtual reality computer classroom so highly touted by various leaders, including some professional educators, is a very, and I do mean a *very*, poor substitute for a well educated human teacher, some good literature, an adequate microscope, sharp dissecting tools, and a bucket of swamp gunk.

The long term value of natural classrooms is simply a given, although it is a constant battle between biology teachers and other kinds of "teachers" to keep this fact in the mind of a public that seems to outgrow its hard-earned history lessons very quickly. But those history lessons are important indeed. First, economies and cultures cannot be sustained over the long term unless the environments that support them are also sustained. Second, economies and cultures are themselves interdependent and it is nearly impossible for cultures to flourish when their supporting economies collapse (although flourishing economies can produce cultures of their own). And third, water and energy are two essential ingredients to modern, and especially industrialized, cultures and economies; without them, the economies collapse and drag their cultures down in the process.

It's a long way, philosophically, from a wetland to the American dream, but the latter rests as surely as a towering natural food pyramid upon knowledge of the natural world, and not only the knowledge that lets us exploit that world, but the knowledge that warns us when we've exploited it too much. There are few better laboratories in which to learn the latter than your local swamp.

Reasons (1) and (2) above are actually explanations for our ecologist's seemingly off-the-wall answer ("Insects are the glue that holds life on Earth together.") to our original question ("What good is a swamp?"). In very many ways, insects *are* the glue that holds natural systems together, and to whatever extent we humans depend on natural systems, that bug glue holds us together, too. Insects are the primary primary consumers; that is, the energy and materials of an ecosystem flow mainly into the mass of insects, spiders, and other arthropods before ending up in the larger predators such as robins and warblers.

Reason (3) explains why ten year ecologists, and conservationists in general, are so enamored with wetlands. Many of these people got exposed early on to natural areas, including swamps, and learned from firsthand experience that pictures and television are a very poor substitute for real mud that breeds both crawlies and questions. Some in the conservation movement, however, have tried to calculate the "good" of wetlands in more mundane terms such as dollars. The results are usually startling. In Audubon Magazine, for example, Ted Williams quotes a University of California study showing that 454,000 acres of California wetlands (all that remains from well over 4 million acres) were worth about $10 billion a year, while the Louisiana coastal marshes contribute $2 billion a year in seafood alone (see Audubon, November, 1996).

These dollar values are usually based on a wetland's role as detoxifier of agricultural chemicals and waste, as well as basis for a local food pyramid resting beneath a commercial fishery. And one needs only to do a Google® search using "Hurricane Katrina" as the search terms to find out in a hurry what the Mississippi Delta coastal wetlands could have been worth to the nation had they not been lost for a variety of reasons, most of them of human origin.

In summary, swamps are good because they are so often at the base of both aquatic and terrestrial food pyramids. But swamps are also vitally important to us as living laboratories, from which our knowledge of how ecosystems work is derived.

17. Why are ecologists such nerds?

*. . . a person who avidly pursues . . .
obscure interests, rather than engaging in
more social or conventional activities.*

—Wikipedia definition of
"nerd"

Ecologists are such nerds for the same reason that a large number of scientists in general are nerds, namely because they don't give a damn about the things most of the population cares about. What *does* most of the population care about? That's a relatively difficult question to answer without stereotyping a whole lot of people, and there are not very many hard data to support any answer I might suggest. However, there are some pretty good data that tell us how members of the general population spend a great deal of their time and money, so we might conclude that most people care about those things. For example, sports is truly big business; athletics—from participation to consumption of team paraphernalia and clothing—touches all our lives from youth to old age, intrudes constantly into our living rooms through TV, and occupies much of our conversation. Similarly, newspapers, magazines, and television programs are filled with advertisements, sex, politics, crime, war, and religion.

However, if you asked a large sample of ecologists, or of scientists in general, where sports, sex, politics, advertisements, and religion are on their priority lists of major concerns, my guess is that the typical answer would be "pretty far down" (if on the list at all). I might be wrong, indeed quite wrong, about sex, but not too far off on politics, advertisements, material goods (money), and religion. Suddenly we begin to see why ecologists are such nerds. They don't care much about the things most of the rest of us care deeply about! And in this sense they are different from us; they make up part of what the philosophers call "the other."

The "other" is a general term for those who are different, whether that difference be skin color, native language, religion, sexual orientation, or opinions about natural resource use. So when we as a group care deeply, indeed often very deeply, about material wealth, politics, sex, sports, and religion, but someone else cares little or nothing about those things, then we tend to think of that someone else as a member of the "other." "Nerd" is just another word for "other."

What *do* ecologists care about? Again, we don't really have a great deal of hard data, but we do have a measure of what they spend most of their time and efforts doing, namely teaching and research. In fact, if most of them had their way, they'd probably spend most of their time doing research in ecology and very little of their time teaching, and even far less, if any at all, on sports, politics, material wealth, and religion. From observing the behavior of numerous ecologists over several decades, I'd venture to say that if given half an opportunity most of them would depart from civilization immediately and take off for some reasonably exotic and biologically challenging part of the world to study plants and animals.

When they arrived at this particular place they most likely would not (and I do mean *not*) play golf, seek out famous restaurants, or go windsurfing, or lie around on the beach soaking up the sun. They might spend a few evenings hanging out in a colorful local bar talking about ecology if they could find someone willing to listen. But in general they'd spend virtually all their waking hours studying plants and animals and the relationships between them. The locals would treat these ecologists with a certain amount of fear, disdain, gratitude (for the small amounts of money spent on food and shelter), or tolerance typically reserved for the deranged, depending on where this particular exotic place might be. The ecologists might, or might not, notice either the locals or the treatment.

Ecologists are nerds also because their minds tend to be filled with thoughts about timeless questions and natural processes that operate on a global scale. In other words, they think a lot about who eats whom, who wins competitions (in the Darwinian sense, not in the human athletics sense), and how real estate is really divided up. What would our society be like if we had no such nerds? To be rather brutally honest, we'd be pretty ignorant. Nerds who think a lot about timeless questions and natural processes are the source of most of our understanding about our relationships with the planet that supports us. Not all of us like that understanding, or want to hear about it, but it's part of our body of knowledge that needs to be considered when we make major decisions that alter the atmosphere or vast areas of the landscape. For if there is one thing the eco-nerds have shown us, it's that we humans do in fact have the collective power to make our planet uninhabitable.

The fact that we have such power may come as a shock to you, but it doesn't come as a shock to any professional scientist who, by virtue of training and research, knows that Earth's resources are limited. Granted, Earth is a very rich planet, with enormous resources of land, air, and water. Nevertheless, those resources are finite, and humans have demonstrated repeatedly that they can exhaust them, or use them to their limits, at least locally. It remains to be determined whether we can exhaust them globally, but few if any eco-nerds doubt that we have that ability.

Why do these people think this way? The answer is: because all of them are reasonably well trained in mathematics and perfectly able to calculate the approximate date at which the growing mass of humanity will (unless some unforeseen events stop us) equal the fixed mass of the Earth. While that date is fairly far off, it's only an eye-blink in geological time, so these scientists, who tend to think in natural terms, consider the date rather imminent. Eco-nerds also know full well that humanity will experience truly massive misery long before our collective mass becomes equal to that of the planet. Most people don't want to hear these predictions, or think about them. And for most of us, it's easy to resolve the stress of unpleasant thoughts by calling the ecologists nerds.

The mental occupation with timeless questions puts ecologists somewhat in the same category as poets, other types of free-thinkers, mystics, and dreamers. Indeed, many scientists could also be so characterized, although our political leaders and business executives would prefer that scientists be people who are working hard to solve practical problems and produce economic miracles. Well, some scientists do these things, occasionally on purpose. But most real scientists are like those stereotypical ecologists who in turn are not all that different from other intellectuals.

Somebody has to study the many phenomena studied by nerds. Why? Because over the long haul, these folks collectively produce things that somebody on Earth needs and wants. And, rarely do nerds start wars, which is probably a fairly important reason to keep them around. [As an aside, wars are generally started by good-ol'-boys who have convinced us that we simply must have whatever the war will produce.]

Now that we've covered the important reasons why ecologists are such nerds, I need to mention a few of the unimportant reasons, beginning with their clothing. A large percentage of male ecologists dress like outright slobs, and some of them seem to take a fair amount of pride in doing so. Do these habits make them bad people? No. But it does make the businessmen and mid-level university administrators disinclined to hang around with ecologists very much, or to treat them with a whole lot of respect in public. When it comes to scientists, businessmen and middle managers seem to prefer people with white lab coats in clean, glassware- and chemical-infested environments with radioactivity warning signs on all the doors.

Should male ecologists be a little more dapper? No. If they suddenly started showing up for work in 3-piece suits and $50 ties, their ecologist friends would probably disown them. I certainly don't have any serious statistics on ecologists' wardrobes, of course, and furthermore, such statistics probably don't exist. So my comments about clothing, and the relationship between clothing and the respect of businessmen and middle managers, are all impressions based on about 50 years of experience. Nevertheless, they're pretty close to target. The female ecologists seem to do a little bit better than the males in terms of clothing selection. No stereotype intended.

The other major unimportant reason why ecologists are such nerds is that most of them are employed as faculty members at colleges and universities. In American society today, intellectuals of all kinds, and especially university professors, are portrayed as nerds, for example in movies and sitcoms. Should we worry about these poor nerdy ecologists? Probably not; they don't do a whole lot of worrying about themselves. Should we worry about what they are telling us about planet Earth and our utilization of it? Probably. In fact, we probably should be paying a great deal of attention to what the scientists in general are telling us, and a good deal less attention to what the politicians, salesmen, athletes, ministers, and good-ol'-boys are telling us. But with that last bit of advice, I've crossed over the line into polemical territory (although not very far). And finally, should we worry about the poor nerdy image ecologists acquire just from being intellectuals? In my opinion, yes, we should worry when our society treats as "the other" those intelligent, highly educated, people who think about timeless questions.

18. Why do scientists argue?

Where there is much desire to learn, there of necessity will be much arguing . . .
—John Milton (*Areopagitica*, 1644)

Scientists argue for several reasons. First, they're humans, and humans argue, period, about all kinds of things from politics to . . . you name it. Second, scientists argue because they tend to have strong minds and solid opinions about fairly complex matters. Put two of these types of people in the same room, especially if they are experts on the same general phenomenon (for example, the weather), and eventually they'll find a way to express different opinions about something important to them both. Third, scientists argue about the best ways to actually conduct a scientific study. In other words, they don't always agree on the best technology to use in order to solve a problem (in this regard, scientists are similar to coaches). Fourth, scientists argue because of what they personally believe to be important, as opposed to unimportant, questions to address (see "strong opinions" above). Finally, and perhaps most important relative to our understanding of ecology, scientists argue because they have different interpretations of the same set of observations. Scientists may argue for many other reasons, too, but these five are certainly enough for any book entitled *Ten Minute Ecologist.*

Most of the reasons listed are characteristic of the world of science in general, not just ecology. Every scientist is quite familiar with arguments, controversies, alternative explanations for phenomena, conclusions that are shown to be false, seemingly trivial observations later shown to be quite important, etc. Every scientist also is familiar with the underlying motivation to argue—from petty personality differences to legitimate technical alternatives for a particu-

121

lar study. To expand on these sources of disagreement, illustrating them with specific examples, would be to write a book on the history of science, or sociology of science, depending on which examples were chosen. The Manhattan Project, our effort to build an atomic weapon during World War II, is a great example of scientific research in which there was much at stake, but in which there also were many disagreements, from personal to technical. For this reason, the Manhattan Project is an excellent case study that has been explored in numerous books.

Both the history and the sociology of science can be exciting and glamorous, again depending on specific examples. The excitement and glamour grow out of the true importance of the science involved. Obviously the Manhattan Project was an exciting endeavor, carried out by people who could be made larger than life because of their intelligence and what was at stake. So much was at stake in the Manhattan Project, and the actual characters were so intelligent and strong willed, that, in addition to the books, more than one feature film has been made about it (the most recent was *Fat Man and Little Boy*, starring Paul Newman).

Most projects, however, are much more mundane than the Manhattan Project. Nobody makes a feature film about some poor graduate student studying the parasites living inside an obscure beetle. And, although the letters DNA are quite familiar to many of us, if for no other reason than their use in the O. J. Simpson trial and other highly publicized events (e.g., an investigation to determine whether Thomas Jefferson actually sired children by a slave woman), and at times are surrounded by high drama, the truth is that the overwhelming majority of people studying DNA are doing it squirreled away in labs where they're working on pretty mundane problems with no screenplay potential. However, you still hear them standing around in the halls arguing about the best way to do a particular experiment. That's because they're scientists.

Before going into some depth on scientific arguments and what they mean to a ten minute ecologist, I need to answer a question that may be in your mind: why would anyone work on a problem that's mundane, unglamorous, with not much at stake? This question is very easy to answer: because many such problems are great, and I do mean truly *great*, training exercises that teach a person how to do a lifetime of meaningful teaching and research as a scientist. (Many such problems are also pretty useless as training exercises, but this fact often gets discovered in a hurry and the problems get dropped; usually.)

Furthermore, a lot of these kinds of problems are truly captivating ones, once a person understands their biological (as opposed to human) significance. So that's why thousands of people work on projects that politicians may not think are very important and thus perhaps don't want to support with tax money. The study of mundane problems, however, is the one activity that probably does more than anything else to ensure that our nation, with its extreme dependence on technology, has an adequate supply of trained scientists and teachers of science.

So much for the arguments of scientists in general. Why do ecologists argue? One of the main reasons ecologists argue is that their observations are not always as unequivocal as those obtained, for example, in a canned freshman chemistry lab experiment. Phenomena such as rainfall, wind velocity, distribution of plants, activity of insects, numbers of birds in an area, etc., are all the stock in trade of ecologists. But these phenomena are not always easily measured precisely, nor are experiments done outside the lab particularly easy to design and carry out. These experiments take a lot of time, patience, physical labor, and sometimes money.

And, to be completely honest, most wild animals are really dumb, especially insects, worms, and snails (= most animals); plants are not always very cooperative, either. Furthermore, the weather, which is about as uncontrollable as natural phenomena get, has as much effect on an ecologist's re-search as it does on a farmer's business. So there are a lot of ways for variation to enter an ecologist's studies, and in science, variation always, and I do mean *always*, increases the potential for serious discussion (= argument) about observations.

The other reason ecologists argue is that in the past few decades their observations have acquired political significance, largely because of predictions derived from those observations. In the vast majority of scientific endeavors, observation and prediction go hand in hand as working tools. "Prediction" in this case means a testable hypothesis. Scientists usually are very comfortable with predictions, including those that are not later supported by observation (data). The latter case is called "rejection of the hypothesis" and it is a highly useful and productive event in the life of most scientists. In fact, real scientists spend most of their research effort trying to reject hypotheses (= truly test their predictions).

But when scientists predict that the Earth's average temperature will rise enough to melt the ice caps, flood Los Angeles, and alter the agricultural economy of nations, then the testing of that hypothesis becomes a rather serious matter of interest to politicians. The testing of that prediction also then becomes a scientific activity vulnerable to variation introduced by theatrics, demagoguery, lies, self interest, and e-motion. So when ecologists suggest causes for global warming, they find themselves involved in politics because of the rather dire consequences of this prediction, should it come true.

Accumulation and interpretation of data, writing of hypotheses and predictions, and suggestions of causality are all typical daily activities of scientists, including ecologists. Politics, including theatrics, demagoguery, lies, and emotion, are not typical daily activities of scientists. Well, at least not most scientists. Nevertheless, when scientists claim to have discovered things that large numbers of people would like to deny exist (e.g. ozone depletion, global warming, etc.), then scientists get dragged into arguments because of these discoveries.

Depending on their discoveries' significance, scientists also may find themselves accused of "not really proving" that something is true or highly likely. In some cases, scientists may discover that people with political agendas have gone looking for, and found, scientists who disagree with other scientists' interpretations, then used that disagreement for political purposes. Then scientific arguments get dragged out of the lab and into the public glare. At that point, the voting public needs to be very perceptive in order to distinguish the science from the politics.

But perhaps the most important reason of all why scientists argue is the fourth one given in the opening paragraph of this chapter, namely their differing opinions on important, vs. unimportant, questions and answers. Like most areas of original intellectual endeavor, including art, music, and literature, science is guided by prevailing paradigms. Paradigms are really collections of legitimate questions and allowable answers. The questions result in part from available technology; the answers are produced by past research. If you're a scientist, the quickest way to be labeled a jerk is to spend your time addressing illegitimate questions with forbidden answers.

For example, most questions about psychic powers are considered illegitimate by the scientific community and answers involving levitation are not allowed. In general, scientists, including ecologists, consider questions on the legitimate list to be relatively important, i.e., these scientists are guided by the paradigms of their discipline. One of the quickest ways to start an argument among scientists, including ecologists, is to promote the importance of some question not on the legitimate list. The only way to end such an argument is to demonstrate, through research, that your previously illegitimate question is in fact an important one. When this happens, pariahs become heroes, and vice versa. Well, maybe it's not quite that dramatic, but to those involved in scientific arguments, these events often seem that dramatic.

In summary, it's important for a ten minute ecologist to remember two things about scientific arguments: (1) they are a regular and normal aspect of science, and (2) scientists don't usually make controversial predictions without some reason. Typically that reason is preliminary data or observations. Because of (2), it's not always a good idea to completely ignore scientists' predictions just because these hypotheses "have not been proven." In many cases, by the time the prediction has been adequately tested ("proven"), the damage has already been done.

19. Do we humans live by the same rules as beetles?

A few strong instincts, and a few plain rules.
—William Wordsworth (*Alas!*
What Boots the Long Labor-
ious Quest? 1815)

Yes, humans and beetles both have little choice but to live by the same rules that govern all life on Earth: we're born (or hatched, in the case of beetles), live for a relatively short period, reproduce if we can, then die. The difference between humans and beetles lies in the activities we engage in during the relatively short period we live. But then the differences between two species of beetles, aside from their structural differences, also lie in the activities they engage in during their relatively short lives. The same things could be said of any two individual beetles, except that their individual lives (as opposed to the lives of their kind) are partly governed by chance events over which they have no control. And the same can be said for any two individual humans, including the bit about chance events over which they have no control. So yes, humans and beetles, in fact *all* of us that occupy planet Earth, live by these same rules.

Humans, like beetles, must find food, water, and shelter in order to live from day to day. This simple rule is obvious to all of us. What's not obvious to all of us, or at least doesn't seem to be so obvious, is that we must obey this rule just like beetles must obey it. The vast majority of human time is spent searching for food, water, and shelter, although we often call this search "work" or "a job." Many of our most pressing human problems are linked to the search for food, water, and shelter. Unemployment, poverty, public education, agricultural politics, the illegal drug trade, illegal immigration, clearing of tropical forests, clear-cutting of timber, and a long list of other phenomena that we read about in the newspaper, are all associated with the search for food, water, and shelter. That last statement might seem like a violation of my claim of neutrality in the eco-polemics wars, but it's not a violation of that claim so much as it is a rather philosophical, and somewhat detached, view of our general situation here on Earth. We take resources from the planet, spending time and effort to do so, and convert those resources into human tissue. And some of us do it more successfully than others. The same could be said for beetles.

Unlike beetles, however, humans spend a fair amount of time searching for "meaning" as well as for food, water, and shelter. For humans as a whole, "meaning" has many definitions, so not surprisingly, the search has just as many outcomes. To most people, this particular search leads them toward religion; for others, it leads to encounters with drugs, alcohol, athletics, art, music, literature, money, or political power, to mention a few ends. Thus because of our evolved traits (massive and complex central nervous system) our lives are governed, in part, by rules that beetles (insofar as we know) don't have to follow (search for "meaning"). Beetles don't have to live by all our rules (speed limits, for example, or clothing in public), but we still have to live by the basic beetle rules involving food, water, and shelter. As a *species* we have to live by the ultimate living organism rule, too: reproduce or die. So aside from this business of "meaning," we're pretty much in the same category as beetles, i.e., animals moving around over the face of Earth looking for food, water, and shelter.

There are about 400,000 known species of beetles, so by inference they find food in about that many different kinds of ways. There are also uncountable billions of beetle individuals out there looking for food, and not infrequently they find it in human cupboards, grain bins, and warehouses. Across the face of Earth, there are easily countable billions of humans (namely about 7 [billion]) out there looking for food, and these approximately seven billion people also obtain food in a remarkable number of ways. If humans have enough money, they typically find their food in a local store or market.

That statement sounds almost trivial and dumb, but what passes for food in a local store or market, and what that "food" goes through prior to arriving at the market, are both highly dependent on climate, geography, and culture. As I've indicated in other chapters, global events such as past continental drift and prevailing winds, tend to dictate agricultural economies. At least to some extent, those same events probably contribute quite a bit to local culture (for example, gauchos and cowboys are independently evolved economic and cultural life styles associated with particular agricultural practices in similar habitats).

Modern transportation and shipping have erased many of the effects local agriculture has on the search for food. One can buy a fast food franchise hamburger in most parts of the world, for example, or fresh ocean fish deep in the heart of a large continent. Prior to invention of the airplane, Sioux Indians sure never picked up a live Maine lobster, but now you can do it any day of the week in Omaha. Conversely, if Formosan aborigines (living on the island we now call Taiwan) made noodles five hundred years ago, they made them out of something other than Kansas wheat (as they do today). And if one delves further back into human history, especially into the Middle Ages and contemporary aboriginal cultures, the relationship between local geography and feeding habits becomes reasonably obvious. Yes, without modern transportation, we seek food in our immediate surroundings, just like beetles. The ten year ecologists who study global energy supplies foresee a time that we may well return to such beetle-like highly localized agricultural economies.

Humans do differ from beetles in one important way, however: people generally don't get eaten by birds and other animals, although once in a while you read about people being attacked by mountain lions or other large carnivores, usually while engaged in some recreational pursuit like camping or horseback riding. But just like beetles, humans get infected with viruses, bacterial, fungi, and various kinds of animal parasites like nematodes and tapeworms. So quite literally we do get eaten by other organisms who by definition live higher on the trophic pyramid than we do (see chapter 7). Humans play host to at least 250 different kinds of animal parasites alone (worms, lice, ticks, fleas, etc.) When viruses, bacteria, and fungi (e.g. athlete's foot) are added, the list gets considerably longer. So most certainly we are prey as well as predator, another rule with which beetles are well acquainted.

The "reproduce or die" rule that applies to all *species* is one of the most interesting of all rules in biology. Why is it so interesting? Because it's the ultimate species-level paradox: everyone has to follow it but there are at least as many ways to follow it as there are species! I say "at least" because with the smarter species (humans, chimpanzees, elephants, whales, and the like) there are many individual variations on the species' theme. Come to think of it, there are individual variations on species' themes in lots of dumber animals, too, for example as when fruit flies of a certain genetic makeup can't sing the right mating song. What happens to fruit flies with the "wrong" genetic makeup? They miss out on a lot of fly sex. And, tragically (yes, tragically; I am being serious for a moment), the most effective way for humans to miss out on a lot of sex is to be born with a genetic makeup that makes them unattractive to other human beings. On the positive side, however (I'm still being serious), human attractiveness is very often quite negotiable, an option that neither fruit flies nor beetles enjoy.

Right now, both humans and many species of beetles (insofar as we know) are reproducing quite nicely thank you, and are in no danger of dying as a species. But we are also clearing the tropical forests at the rate of about a hundred acres a minute, and that's where most beetles, and indeed most plants and animals of all kinds, live. So in order to satisfy our beetle-like rules (find food, shelter, water, and a mate or mates), we're eliminating the places where literally thousands of species of beetles are doing, or at least trying to do, the same thing. Indeed, instead of thousands, the numbers of beetle species being deprived of food, water, shelter, and mate(s) may easily number in the hundreds of thousands.

How do we arrive at this conclusion? Easy. By counting the rate at which new species are being discovered in known areas we can estimate the number of species yet to be discovered. It remains to be seen whether we ultimately will have to live by beetle rules in the tropics, i.e. that when the forests are gone, we, like the beetles, will wander the charred ground looking unsuccessful for food, water, and shelter. We'll probably find mates. We always seem to find someone to mate with, no matter how miserable our circumstances.

So, in summary, yes indeed, we do live by the same rules as beetles, as well as some rules of our own, but technology has given us the power to forestall the application of a few of those rules. It remains to be seen whether we can avoid living like beetles forever, and indeed "environmental controversies" are often a product of differing opinions on just how long "forever" is.

20. Did God make the Earth in seven days?

We must build a new world, a far better world...

—Harry Truman (April 23, 1945)

No, God did not make the Earth in seven days. If a God made the Earth, He/She/It constructed the planet over a four to five billion year period, using the full suite of natural processes that humans have discovered over the past two or more millennia. These processes include all of the astronomical, geological, atmospheric, and biological ones mentioned in any halfway decent freshman science text plus perhaps some yet to be discovered ones. So, no, God did not make the Earth in seven days. You can believe anything you want about the origin of Earth, or the origin of all the surface features on it including ourselves, but the vast preponderance of scientific evidence indicates the Earth is about 4.6 billion years old and all of us evolved on it.

On the other hand, there is no evidence whatsoever for the existence of any other planets *just like Earth* in the universe. Although there is plenty of evidence that planets exist—astronomers are finding them more and more frequently—and some well founded speculation about the potential number of inhabited planets in the Universe, insofar as we know for certain we're alone and unique. This knowledge is enough of a reason for wonder and spirituality, but not enough of a reason to dismiss out of hand the past two millennia of scientific observation about Earth and its origins.

Why do ten minute ecologists need to know something about the origin of Earth and its inhabitants? Because geophysical and biological processes have been at work here for several billion years, and they are still at work to produce the environment in which we, and all other of Earth's inhabitants, live. These processes range from the local to the cosmic, from the drifting of continents and changing global climate, to the falling of meteorites, from volcanoes, floods and earthquakes to sunspot cycles, from algal blooms to the burning of fossil fuels.

Knowledge of these processes in no way diminishes the wonder anyone must have for the complexity and beauty of Earth. Such knowledge, however, does tend to emphasize both our uniqueness, and more significantly, our common interest in the health of our planet. That common interest demands that our collective political, economic, and agricultural actions maximize the Earth's ability to recover from the use we make of Her. This interest also demands that we minimize our own impact on Earth's resources of land, plants, animals, minerals, water, and air, which is, of course, the same thing as maximizing the planet's ability to recover from our actions. As any ecologist, myself included, will tell you, political and economic stability, standard of living, and cultural richness all depend on long term sustainable use of our environment. This last statement is not a polemical one so much as an observed fact.

How *was* Earth made? Available evidence indicates it condensed from a cloud of gas and other particles not long after the sun itself was formed from a larger cloud of gas (or, perhaps, as a consequence of the sun's formation). Although, of course, no one has ever watched the complete formation of a star and its planets from beginning to end, modern astronomers have studied star formation and evolution extensively using a variety of means and technology, including space telescopes such as the Hubble. No one has ever literally watched a chemical reaction take place, either, although, as in the case of planets and galaxies, lots of us have watched the results of chemical reactions.

Astronomers also have behaved as most scientists do by developing theories, designing instruments to help test these theories, and making the measurements necessary to test their predictions and hypotheses. As a result of all this activity, we know a great deal about the Sun, the Solar System, the Milky Way galaxy, and the rest of the universe. Their conclusion, so far, is that we (you and I and Earth) began in a gas cloud about 4.6 billion years ago.

Geophysicists, likewise, have behaved about like other scientists in developing theories, making predictions based on these theories, then designing studies and instruments to test the predictions. Geophysicists tell us that the Earth's crust is made from a number of plates that "float" on the molten rock below. These plates not only carry the continents, but they also move around, drift apart or together, and bump into one another, their edges slipping beneath or riding up over other edges, all of which processes ultimately produce earthquakes and volcanoes. The moving of plates also produces vast changes in the climate, flora, and fauna of a region, but over long periods of time. When a continent drifts, plants, animals and microbes go along for the ride, sometimes to parts of Earth that, because of latitude and a tilted axis of rotation may not be overly hospitable. Thus when a continent drifts from the tropics to the poles, organisms on that continent must ride along and adapt, in the evolutionary sense, or become extinct.

But some species, such as barn owls and barn swallows, are quite capable, even today, of traveling quickly between continents, so that the drifting doesn't split and isolate barn swallow and barn owl populations in quite the same way that the same process tends to break up populations of freshwater fish, for example. But for most of Earth's history and most of Earth's organisms, continental drift has been a major factor in producing our present day diversity and distribution of plants and animals.

Cichlid fishes, for example, which are members of a freshwater family, evolved on a continent that later split into Africa and South America. As a result of the split and subsequent drift, cichlid fishes are now found naturally in those two widely separated places. This kind of distribution is typical of the observations used to support theories of continental drift early in this century. In the case of Africa and South America, of course, the coastline profiles also suggested a breakup far in the prehistoric past. Modern geophysics has since confirmed what the turn-of-the-[last] century biogeographers suspected, namely that the two continents were once joined.

One the most significant of the ancient geological events is known as the "breakup of Pangaea." Pangaea is the name of a single continent that was formed about four hundred million years ago from the collision of all the rest of the world's ancient continents, including Gondwana and Laurentia. Dinosaurs flourished on Pangaea. About 135 million years ago, Pangaea started to break apart, and over the next hundred million years broke apart into several other continents that ultimately drifted to their present positions. We now call these continents Africa, Eurasia, North America, South America, Antarctica, and Australia. Thus the present distribution of plants and animals is partly a result of having had ancestors who lived on Pangaea, and subsequently rode a part of Pangaea to its present position while evolving along the way.

Today we can see the results of all these seemingly (to us) slow geological and biological processes at work. For example, the marsupials (pouched mammals such as kangaroos) became geographically isolated from the rest of the world when what is now Australia broke away from the rest of Earth's land masses, and remained so; this isolation was a major factor in their evolutionary history. When we go to the zoo and marvel at the wondrous Australian fauna, we're actually reacting intellectually to the geological and evolutionary events of the past fifty million years. Those events not only produced kangaroos, wallabies, koalas, and all their relatives, but also kept them separated from the pre-dogs and pre-cats that eventually evolved into jackals, lions, and tigers.

In the past three hundred years, the British colonized Australia, starting with the transportation of criminals, cats, dogs, and rats to the continent. That colonization was the beginning of our biological knowledge of the isolated continent. Had the Australian aborigines sent their criminals to North America by the shipload then followed as administrators, tourists, and businessmen, then there might well be outback zoos filled with marvelous exotic animals such as wolves, bison, wild turkeys, and deer. But that didn't happen. Instead, the British colonized Australia and we are the ones who are fascinated by the exotic fauna. And a ten minute ecologist at the zoo now knows that we're fascinated by kangaroos because they are a product of geophysical and evolutionary events that occurred over the past millions of years, far away from our familiar flora and fauna.

Evolution of Australian fauna and flora, extinction of the dinosaurs, distribution of cichlid fishes, and distribution of land masses and oceans, are all examples of "the way the universe works." Scientists seek explanations for "the way the universe works," i.e. natural phenomena. We now know that the phrase "natural phenomena" includes everything from the formation of galaxies to the innermost functions of a cell. And we also know that scientists study the universe at all these levels, from the grand and cosmic to the detailed and microscopic. Ecologists have focused their attention on mechanisms that produce and maintain the diversity of life on Earth, with special emphasis on interactions between organisms—including humans—and their environments. There is a compelling reason for such study, and that is the obligate relationship we all hold with the only planet in the universe actually known to support life.

My hope is that this small volume will serve to inform its readers of the many facets of that relationship and that such information will be of eventual use to us all. As we have learned so often in the past, ignorance is far more dangerous than understanding. But *Ten Minute Ecologist* is only an introduction to the vast body of knowledge about our planet. Wading through the suggested readings (see appendix) will be a little like wading through a swamp in that you'll find everything in that list from organisms to philosophy. If *Ten Minute Ecologist* has done its job, then that appendix will share one other attribute of swamps, namely the fascination they hold for those who've ventured into them out of curiosity, and like an unrepentant naturalist, you'll go looking for more.

Glossary

Abiotic factors Non-living components of Earth—for example, wind, rocks, water, light, various chemical elements.

Algae Several groups of photosynthetic organisms, many consisting of only a single cell, most living in water, and all lacking the tissues that transport fluids in non-algae plants such as familiar neighborhood trees.

Arithmetic growth Growth in which change in numbers or size is a constant regardless of previous numbers or size, as in a payroll deduction applied to a savings account.

Atmosphere The layers of gasses that surround Earth.

Asexual reproduction Multiplication of organisms (production of new individuals) without sex. Many if not most single-celled organisms, as well as fungi, and some worms, multiply asexually.

Autotrophic Literally, "self feeding," but in biology, referring to plants that use sunlight energy to build carbohydrate molecules from carbon dioxide and water.

Bacteria Single-celled, microscopic organisms without a membrane surrounding their nuclear material (their genetic information, their DNA) and lacking the well-formed membrane-bound structures typical of plant and animal cells.

Baleen whales Whales that strain their food from the ocean using sheets of tissue that hang down from the roof of their mouth and function as a strainer.

Beetles Exceedingly numerous insects of the order Coleoptera, which typically possess front wings hardened into covers.

Biodiversity The variety of living organisms; the number of species.

Biome A type of environment distinguished by its characteristic temperature, available moisture, and dominant vegetation.

Biotic factors The living organisms of Earth, as well as non-living products or living organisms, including, for example, shells, feces, skeletons of dead animals, fallen leaves, and burrows.

Blubber A layer of fatty tissue beneath the skin of marine mammals such as whales and walruses.

Boreal Northern, particularly the forested regions of the high latitudes in the northern hemisphere.

Brackish (water) Water that is partly salty, usually resulting from the mixing of seawater and freshwater, as at the mouth of a river that empties into the ocean.

Carnivore Meat-eating animal.

Class One of the taxonomic groups in biological classification, included in a phylum and containing one or more subordinate groups called orders.

Climax vegetation The stable, dominant type of vegetation that occupies a particular area and is determined largely by temperature and available moisture.

Community All the species that occupy a particular habitat and presumably interact with one another in some way. An ecologist may selectively refer to some of these species, for example, as in "plant community" or "parasite community" (living in and on other animals).

Competition An ecological interaction in which organisms attempt to obtain, or occupy, a limited shared resource such as space, water, or food.

Conifer Evergreen, cone-bearing trees such as spruce, fir, and pine.

Consumer An organism that eats other organisms, typically another species, thereby obtaining energy and nutrients.

Continental Drift A geological process in which the Earth's continents move across the face of Earth because the crustal plates upon which these continents rest are themselves moving.

Crustaceans Arthropods such as crabs, crayfish, shrimp, lobsters, and tens of thousands of related forms, with two pairs of antennae and appendages with two branches.

Decomposers Organisms, such as fungi and certain bacteria, that break down biological materials. Decomposers digest dead plants and animals.

Describer The individual who is given credit for formally describing a new species in a publication.

Diatom One of the types of algae, characterized by structurally intricate shells made of silicon.

Diversity index A mathematical expression that summarizes the relative diversity of a community, taking into account both the numbers of species and their respective numbers of individuals.

Division A large, relatively inclusive, taxonomic group used by botanists, equivalent to a phylum (used by zoologists).

DNA The large, coiled, molecule that contains genetic information by virtue of the sequence of its parts (rather analogous to a sentence containing information because of the sequence of its letters).

Earthworms Segmented worms of the phylum Annelida, class Oligochaeta, that dwell in the soil.

Ecological niche A "place" in nature occupied by a particular species. This "place" consists of all the resources the species needs in order to survive.

Ecologist A scientist who studies the interactions between organisms and their environments.

Ecology The study of interactions between organisms and their environments, with the intent of discovering how the numbers and distribution of organisms are maintained and controlled.

Ecosystem All the organisms, their interactions, and the environment that supports them, usually within a large, distinctly recognizable area such as a prairie, desert, or forest.

Endangered species A species with a population so small that the species is in danger of becoming extinct.

Enzymes Proteins that aid in carrying out chemical processes such as digestion.

Epiphyte A plant that typically grows on another plant and obtains its moisture from the air.

Evolution Genetic changes leading to the formation of new species that occur in populations of organisms over extended periods of time. Evolution is the central unifying theme of biology.

Exponential growth Growth in which change in numbers or size is a percentage of the previous number or size, as in compound interest applied to a savings account.

Extant Still living and present on Earth, as a species or other taxonomic group.

Extinct No longer present as a species on Earth.

Food chain A linked series of feeding relationships in which organisms feed on other organisms but are in turn fed on by still other organisms.

Food web A group of linked and intercrossing food chains, with opportunistic feeding perhaps at several trophic levels (see **trophic levels** below).

Fossil fuel Coal, oil, and natural gas, all derived from algae and plants that grew millions of years ago.

Fungi Mushrooms, toadstools, and similar smaller organisms that require living or once-living materials as food, function mainly as decomposers, and reproduce both sexually and asexually.

Genus (pl. genera) A category in the biological classification scheme—more inclusive than the species, less inclusive than the family. A genus contains one or more species.

Geophagy The practice of eating dirt (on purpose). Also called chthonophagia.

Glaciation The advance of glaciers to cover parts of Earth.

Greenhouse gasses Several gasses that accumulate in the upper atmosphere and prevent heat loss from Earth.

Guild A group of species competing for the same resource.

Habitat The general type of environment occupied by a species—for example stream, lake, forest, grassland, ocean, tide pool.

Herbivore An animal that eats plants; a primary consumer.

Heterogeneity Literally, "the quality of being different." In the biological sense, heterogeneity refers to the extent to which a group or habitat contains a variety of different members or components.

Hierarchy In biology, a classification system with groups that are increasingly inclusive, as less inclusive groups existing only as components of the more inclusive groups.

Higher taxa The more inclusive taxonomic groups such as phyla, divisions, classes, and orders.

Homogeneity Literally, "the quality of being the same." In the biological sense, homogeneity refers to the extent to which a group or habitat is characterized by sameness.

Interbreed To reproduce sexually with one another. Members of the same species interbreed by definition;

members of different species sometimes successfully interbreed.

Interspecific Between different species.

Intraspecific Between members of the same species

Krill Small shrimp-like crustaceans that feed mainly on phytoplankton and occur in very large numbers in the ocean.

Lichens Organisms consisting of a species of fungus and a species of algae living in a symbiotic relationship.

Loam A relatively loose soil type with organic material, sand, and clay.

Loess Yellowish, loam-like soil generally considered to be originally deposited by the wind.

Lower taxa The less inclusive taxonomic groups such as genera and species.

Membrane In reference to cells, a double layer of lipid ("fat") molecules, along with various proteins, cholesterol, and carbohydrates, that separates the cell from its environment. Membranes also enclose other cell structures, which are then termed "membrane bound."

Microbe A microscopic organism—usually bacteria, some fungi, and viruses.

Microorganism A broader, more inclusive, category than microbe—usually microscopic plants and animals as well as bacteria and fungi.

Moraine A ridge of rock left behind when a glacier melts.

Mutation A heritable, spontaneous, change in the genetic makeup of an organism.

Negative feedback system Any process in which the products make the process itself operate more slowly or less extensively. **Positive feedback systems** operate in the opposite manner (products accelerate the process).

Nerd A person the general public considers strange, often preoccupied with subjects such as math, science, ecology, and computers.

Niche *See* ecological niche.

Nitrogen One of the chemical elements; the major atmospheric gas on Earth.

Order One of the taxonomic groups in a biological classification. An order is a member of a more inclusive **class**, and contains one or more families.

Organic Derived from living organisms such as plants, animals, bacteria, and fungi.

Organism A living being or entity.

Oxygen One of the chemical elements; the second most common atmospheric gas on Earth; a gas required for life as we know it.

Ozone A gas made of three oxygen atoms. Ozone in the upper atmosphere blocks ultraviolet radiation from falling on Earth.

Paradigm A concept or theory that guides our thinking, mainly be establishing a range of legitimate questions and legitimate answers.

Parasite An organism that can exist only by living in a close, dependent, sometimes harmful relationship with another, unrelated, organism.

Parent rock The rock that wears down to produce a particular soil.

Particle-size distribution The relative numbers of soil particles, or various sizes, in a sample.

Pharmaceutical Referring to a chemical substance with medicinal properties, a drug used to treat disease.

Pheromones "Odorless" molecules that are released by animals and in turn stimulate other members of the species to behave in certain ways usually associated with sex.

Phloem One of the plant tissues, made of tube-like cells, that function to transport fluids and sugars.

Photosynthesis The biochemical process by which green plants capture the sun's energy and use that energy to build complex molecules such as sugars.

Phylum A taxonomic group of animals, very inclusive, and containing all the species that possess a similar general body plan.

Phytoplankton Relatively small, photosynthesizing, organisms—for example floating algae—that occupy the open water in lakes and in the ocean.

Pioneer An organism that is among the first to occupy a disturbed area. Pioneer plants have properties such as small windblown seeds that aid them in finding such disturbances.

Plankton Relatively small plants and animals that live suspended in water, either in the ocean or in bodies of freshwater.

Pleistocene A period in Earth's history from about 1.7 million years ago to 10,000 years ago, characterized by advancing and retreating glaciers.

Population The number of commonly and freely interbreeding individuals of a species. Species can occur in more than one population, and in this case, interbreeding between the two or more populations may not be particularly free and common.

Positive feedback system Any process in which the products make the process itself operate faster or more extensively. **Negative feedback systems** operate in the opposite manner (products inhibit the process).

Producers Green plants that capture the sun's energy and use that energy, plus simple molecules, to build complex molecules that in turn can be used by other organisms.

146

Reef A rock-like deposit occurring in warm, shallow, seas and consisting mainly of the skeleton produced by marine animals of the phylum Cnidaria (corals and their relatives).

Reproductive isolation A condition in which groups of organisms, of the same or different species, are unable to successfully interbreed with one another for a variety of reasons.

Rotifers Microscopic animals occurring in water and moist soils, and characterized by an external cuticle made of telescoping rings, a two-lobed rear end, and bands of cilia at the front end.

Seed bank The seeds that are buried in soil and able to germinate if moved to the proper depth and provided with adequate moisture.

Species (pl. species) The fundamental unit of biological classification. A species is a kind of organism, typically distinguished structurally or biochemically (especially in the case of bacteria), usually incapable of interbreeding with other species, and considered to have evolved from a single ancestral species.

Succession A phenomenon in which, over time, a series of species, each with differing characteristics, dominates a disturbed area and are subsequently replaced by other species.

Symbionts Organisms that live in and on other organisms that are not of the same species. Parasites are symbionts, as are the algae that provide energy for reef-building corals.

Systematics The science of classification as applied to evolutionary relationships.

Tardigrades Microscopic animals, also called water bears because of their superficial resemblance to teddy bears. Tardigrades often live in moist soil.

Taxonomy The science of classification. Taxonomists classify organisms and study the basis for such classification.

Tide pool A pool of ocean water left along the shore when the tide goes out.

Trade winds Relatively constant winds that blow mainly over the ocean from northeast or southeast and toward the equator.

Transpiration A process by which plants lose water by means of evaporation through stems and leaves.

Trophic Of or relating to feeding.

Trophic level A measure of the number of times energy is transformed before it ends up stored in an organism. This number of transformations is determined by the number of steps (links) in a food chain.

Trophic pyramid A picture of the energy loss that occurs when energy is transformed between tropic levels. This picture shows that most of the energy stored at one trophic level—for example, as producer green plants—is lost when the plants are converted to primary consumer herbivores. Thus the top levels of this pyramid are smaller than the bottom layers.

Trophic relationship A relationship between organisms, usually between different species, characterized by feeding. In a trophic relationship, one species eats another.

Tropics Earth's warm regions between the Tropic of Cancer and the Tropic of Capricorn.

Undescribed species A species that has not been formally recognized by the scientific community because it has not been described according to the rules of scientific nomenclature in a publication.

Vagility The inherent ability of an organism to disperse geographically.

Virus A parasitic microbe typically consisting of genetic information covered with a protein coat; viruses exist and reproduce only by infecting other organisms.

Xylem One of the plant tissues that transport water and dissolved minerals and provide structural support to plant stems.

Zooplankton Relatively small animals that occupy open water in lakes and in the ocean.

Acknowledgments

I would like to thank the various people who helped with this book project. Scott and Elizabeth Snyder both read parts of the early first edition manuscript and gave their comments and encouragement. Scott was a doctoral student in my lab at the time, and Liz was a middle school science teacher in a rural Nebraska community. Some members of the Nebraska Nature Conservancy board of trustees also read some of the first chapters and provided comments. I greatly appreciate the education received as a result of my service on that board of trustees, an education provided by the Nebraska Nature Conservancy staff, members of the board, and various individuals associated with the national organization of TNC. Cara Stanko and Kayleen Karnopp, students who worked in my office, helped with library research. Jessie Ebers, an honors student in my lab and a recent University of Nebraska graduate, served as reader for this second edition and I am very grateful for her work, both as a researcher and an editor.

I also need to thank St. Martin's Press for publishing the first edition, and my agent Jane Dystel for retrieving the rights to that the second edition could become available to the public. Particular thanks are due Scott Slovic, director of the Center for Environmental Arts and Humanities during the 1990s—a remarkable program designed to help us articulate our relationships with the planet—at the University of Nevada Reno, for supplying additional reading suggestions.

The reference on tree density in tropics is: Pennisi, E. 1994. Tallying the tropics: seeing the forest through the trees. *Science News* 145:362-366. Many of the epigraphs are from *Familiar Quotations: A collection of passages, phrases and proverbs traced to their sources in ancient and modern literature*, 15th Ed, Emily Morrison Beck, Ed. (aka *Bartlett's Quotations*), a truly wonderful book. The rest of the biological information in this book can be found in various forms in virtually all university level introductory biology and ecology textbooks. *Ten Minute Ecologist* is simply a way for people who would never encounter either such textbook, or encountered them unpleasantly in the far distant past, to reacquaint themselves with some of the principles that guide our existence here on this incredible planet Earth.

Suggested Readings:

The following list is intended to be only an introduction to environmental literature that might be available in bookstores and libraries. Most of these works are relatively easy to under-stand and provide information that will put the information in *Ten Minute Ecologist* into a broader context. Anyone who undertakes a serious program of ecological self education will immediately discover a rich and voluminous literature on nature, including works that may be exceedingly useful but are not on this list. Eventually, however, such an explorer will encounter some of the more technical material, including modern texts. My advice is to not be intimidated. It's important that we all try to become better educated in the ecological sense, and the acts of reading, exploring the available literature and internet resources, and talking about ecological issues are all vital to both our long term well being and the solution of immediate political and economic problems involving natural resources.

Abbey, E. 1968. *Desert solitaire: a season in the wilderness.* Ballantine Books, New York. 303p.

Akin, W. E. 1990. *Global Patterns: Climate, Vegetation, and Soils.* University of Oklahoma Press, Norman, OK. 370p.

Barry, J. M. 1997. *Rising tide: the great Mississippi flood of 1927 and how it changed America.* Touchstone; Simon and Schuster, New York. 524p.

Bryce, R. 2010. *Power hungry: the myths of "green" energy and the real fuels of the future.* Public Affairs, New York. 394p.

Buchmann, S. L., and G. Nabhan. 1996. *The forgotten pollinators.* Island Press, Washington, D. C. 292p.

Campbell, D. G. 1992. *The crystal desert: summers in Antarctica.* Houghton Mifflin Company, Boston, MA. 308p.

Carson, R. 1951. *The sea around us*. Oxford University Press, New York. 230p.

Carson, R. 1962. *Silent spring*. Houghton Mifflin Company, Boston, MA. 368p.

Davidson, J. W., and J. Rugge. 1988. *Great heart: the history of a Labrador adventure*. Viking, New York. 385p.

Dessler, A. E. and E. A. Parson. 2006. *The science and politics of global climate change: a guide to the debate*. Cambridge University Press, Cambridge, UK. 200p.

Diamond, J. 1997. *Guns, germs, and steel: the fates of human societies*. W. W. Norton & Company, New York. 480p.

Dillard, A. 1974. *Pilgrim at Tinker Creek*. Harpers Magazine Press, New York. 271p.

Dörner, D. 1996 (American edition). *The logic of failure: Why things go wrong and what we can do to make them right*. Metropolitan Books, Henry Holt and Company, New York. 222p.

Dowie, M. 1995. *Losing ground: American environmentalism at the close of the Twentieth Century*. The MIT Press, Cambridge, MA. 317p.

Eiseley, L. 1946. *The immense journey*. Random House, New York. 210p.

Eiseley, L. 1964. *The unexpected universe*. Harcourt Brace Jovanovich, New York. 239.

Farb, P. 1968. *Man's rise to civilization*. Dutton, New York. 332p.

Farb, P. 1978. *Humankind*. Houghton Mifflin Company, Boston. 528p.

Freudenburg, W. R., R. Gramling, S. Laska, and K. T. Erikson. 2009. *Catastrophe in the making: the engineering of Katrina and the disasters of tomorrow*. Island Press, Shearwater Books, Washington, D. C. 209p.

Gore, A. 2006. *An inconvenient truth: the planetary emergency of global warming and what we can do about it.* Rodale Press, Emmaus, Pennsylvania. 325p.

Gould, S. J. 1989. *Wonderful life: the Burgess Shale and the nature of history.* W. W. Norton and Company, New York. 347p.

Gruchow, P. 1988. *The necessity of empty places.* St. Martin's Press, Inc., New York. 291p.

Hay, J. and P. Farb. 1966. *The Atlantic shore: human and natural history from Long Island to Labrador.* Harper and Row, New York. 246p.

Hertsgaard, M. 1998. *Earth odyssey: around the world in search of our environmental future.* Broadway Books, New York. 372p.

Ives, J. D., B. R. Graham, and D. K . Alford. 1974. *Arctic and alpine environments.* Harper and Row, Publishers. New York. 999p.

Janovy, J. Jr. 1978. *Keith county journal.* St. Martin's Press, Inc., New York. 210p.

Janovy, J. Jr. 1980. *Yellowlegs.* St. Martin's Press, Inc., New York. 192p.

Janovy, J. Jr. 1981. *Back in Keith County.* St. Martin's Press, Inc., New York. 179p.

Janovy, J. Jr. 1985. *On becoming a biologist.* Harper and Row, Publishers, New York. 160p.

Janovy, J. Jr. 1992. *Vermilion sea: a naturalist's journey in Baja California.* Houghton Mifflin Company, Boston. 226p.

Janovy, J. Jr. 2011. *Intelligent Designer: Evolution for Politicians.* Trade paperback from CreateSpace.com; e-book from Kindle, Nook, and Sony Reader.

Janovy, J. Jr. 1994. *Dunwoody Pond: reflections on the high plains wetlands and the cultivation of naturalists.* St. Martin's Press, Inc., New York. 288p.

Kagawa, F., and D. Selby (eds). 2010. *Education and climate change: living and learning in interesting times.* Routledge, New York. 259p.

Kaplan, R. D. 1998. *An empire wilderness: travels into America's future.* Random House, New York. 393p.

Krebs, C. J. 1988. *The message of ecology.* Harper Collins, Publishers, Inc., New York. 195p.

Krutch, J. W. 1969. *The best nature writing of Joseph Wood Krutch.* William Morrow and Company, Inc. New York. 384p.

Leopold, A. 1949. *A Sand County almanac.* Oxford University Press, New York.

Lopez, B. 1986. *Arctic dreams: imagination and desire in a northern landscape.* Bantam, New York.

Madsen, J. 1982. *Where the sky began: land of the tallgrass prairie.* Houghton Mifflin Company, Boston. 321p.

Matthiessen, P. 1991. *African silences.* Random House, New York. 225p.

McNeill, W. H. 1977. *Plagues and peoples.* Doubleday, New York. 365p.

McPhee, J. 1980. *Basin and range.* Farrar, Straus, and Giroux, New York. 216p.

McPhee, J. 1982. *In suspect terrain.* Farrar, Straus, and Giroux, New York. 210p.

McPhee, J. 1986. *Rising from the plains.* Farrar, Straus, and Giroux, New York. 214p.

McPhee, J. 1989. *The control of nature.* Farrar, Straus and Giroux, New York. 272p.

McPhee, J. 1993. *Assembling California.* Farrar, Straus, and Giroux, New York. 304p.

Mayr, E. 1982. *The growth of biological thought.* Belknap Press (Harvard University Press), Cambridge, MA. 974p.

Mayr, E. 1988. *Toward a new philosophy of biology: Observations of an evolutionist*. Belknap Press (Harvard University Press), Cambridge, MA. 564p.

Nabhan, G. 1993. *Counting sheep: twenty ways of seeing desert bighorn*. University of Arizona Press, Tucson, Arizona. 261p.

Nishizawa, T., and J. I. Uitto (Eds). 1995. *The fragile tropics of Latin America: sustainable management of changing environments*. United Nations University Press, New York. 325p.

Owen, D. F. 1976. *Animal ecology in tropical Africa*. W. H. Freeman, San Francisco, CA. 122p.

Perry, R. 1973. *The Polar Worlds*. Taplinger Publishing Company, New York. 316p.

Quammen, D. 1996. *The song of the dodo: island biogeography in an age of extinctions*. Scribner and Sons, New York, NY. 702p.

Reisner, Marc 1986. *Cadillac desert: the American west and its disappearing water*. Viking, New York. 582p.

Remmert, H. 1980. *Arctic animal ecology*. Springer-Verlag, Berlin. 250p.

Ricketts, E. F., J. Calvin, and J. W. Hedgpeth (revised by D. W. Phillips). 1985. *Between Pacific tides* (5th ed.). Stanford University Press, Stanford, CA. 652p.

Schell, J. 1982. *The fate of the Earth*. A. A. Knopf, New York. 244p.

Slovic, S. H., and T. F. Dixon (eds). 1993. *Being in the world: an environmental reader for writers*. Macmillan Publishing Company, New York. 726p.

Smith, R., and B. Lourie. 2009. *Slow death by rubber duck: the secret danger of everyday things*. Counterpoint, Berkeley, California. 328p.

Spray, S. L., and M. D. Moran (eds). 2006. *Tropical deforestation*. Rowman & Littlefield, Lantham, Maryland. 193p.

Steinbeck, J. 1941. *The log from the* Sea of Cortez. Viking, New York, 336p.

Sutton, G. M. 1961. *Iceland summer*. University of Oklahoma Press, Norman, OK. 253p.

Sutton, G. M. 1985. *Eskimo year* (2nd edition). University of Oklahoma Press, Norman, OK. 321p (originally published in 1934 by The Macmillan Company).

Terborgh, J. 1992. *Diversity and the tropical rain forest*. Scientific American Library (W. H. Freeman), New York. 242p.

Thomas, L. 1974. *Lives of a cell: notes of a biology watcher*. The Viking Press, New York. 153p.

Trudge, C. 2005. *The tree*. Three Rivers Press (Random House), New York. 459p.

Vandermeer, J., I. Perfecto, and M. J. Jody. 1995. *Breakfast of biodiversity: the truth about rain forest destruction*. Oakland California Institute for Food and Development Policy. 185p.

Ward, P. D. 2010. *The flooded Earth: our future in a world without ice caps*. Basic Books, New York. 261p.

Weisman, A. 2007. *The world without us*. St. Martin's Press, New York. 324p.

Wells, S. 2010. *Pandora's seed: the unforeseen cost of civilization*. Random House, New York. 230p.

Wilson, E. O. 1992. *The diversity of life*. W. W. Norton and Company, New York. 424p.

Wilson, E. O. 1994. *Naturalist*. Island Press, Washington, D. C. 380p.

Young, L. B. 1977. *Earth's aura*. A. O. Knopf, New York, NY. 305p.

Zimmerman, J. L. 1990. *Cheyenne Bottoms: wetland in jeopardy*. University of Kansas Press, Lawrence. 197p.

Zinsser, H. 1934. *Rats, lice and history*. Atlantic/Little Brown, Boston. 301p.

Zwinger, A. 1975. *Run, river, run: a naturalist's journey down one of the great rivers of the west.* Harper and Row, New York. 317p.

Zwinger, A. and B. E. Willard. 1972. *Land above the trees: a guide to American alpine tundra.* Harper and Row, New York. 489p.

The Author:

John Janovy, Jr. received his PhD from the University of Oklahoma in 1965 and has been a faculty member at the University of Nebraska Lincoln since 1966. His research interest is parasitology, with particular focus on ecology and life cycles. He is now retired, but when active held the Paula and D. B. Varner Distinguished Professorship at UNL. He has been Director of the Cedar Point Biological Station, Interim Director of the University of Nebraska State Museum twice, an Assistant Dean of Arts and Sciences, and has recently completed a 6-year term as secretary-treasurer (chief business officer) of the American Society of Parasitologists. His scholarly and creative accomplishments include almost 100 scientific papers and book chapters; 18 books; the screenplay for the televised version of *Keith County Journal* (Nebraska Public Television); and, numerous popular articles. His teaching experiences include almost continuous service in large-enrollment freshman biology courses, Field Parasitology at CPBS, Invertebrate Zoology, Parasitology, Organismic Biology/Biodiversity, and numerous honors seminars. He has supervised 18 MS, 14 PhD students, and approximately 100 undergraduate researchers (10 Howard Hughes scholars). His honors include the University of Nebraska Distinguished Teaching Award (1970), State of Nebraska Pioneer Award (1983), University Honors Program Master Lecturer (1986), American Health magazine book award (1987, for *Fields of Friendly Strife*), University of Nebraska Outstanding Research and Creativity Award (1998), The Nature Conservancy Hero recognition (2000), Nebraska Library Association Mari Sandoz Award (2002), and the American Society of Parasitologists Clark P. Read Mentorship Award (2003).

www.ingramcontent.com/pod-product-compliance
Lightning Source LLC
Chambersburg PA
CBHW031322290526
45784CB00014B/782